In situ generated β-$Yb_2Si_2O_7$ environmental barrier coatings to protect non-oxide silicon-based ceramics in gas turbines

In situ generated β-$Yb_2Si_2O_7$ environmental barrier coatings to protect non-oxide silicon-based ceramics in gas turbines

Von der Fakultät für Ingenieurwissenschaften
der Universität Bayreuth
zur Erlangung der Würde eines
Doktor-Ingenieurs (Dr.-Ing.)
genehmigte Dissertation

von

Eng. Mateus Lenz Leite
aus
Itajaí, Brasilien

Erstgutachter:	Prof. Dr.-Ing. Walter Krenkel
Zweitgutachter:	Prof. Dr.-Ing. Ralf Moos
Tag der mündlichen Prüfung:	08. Dezember 2021

Lehrstuhl Keramische Werkstoffe
Universität Bayreuth
2021

Bibliografische Information der Deutschen Nationalbibliothek
Die Deutsche Nationalbibliothek verzeichnet diese Publikation in der
Deutschen Nationalbibliografie; detaillierte bibliographische Daten sind im Internet
über http://dnb.d-nb.de abrufbar.
1. Aufl. - Göttingen: Cuvillier, 2022
Zugl.: Bayreuth, Univ., Diss., 2021

Nonnenstieg 8, 37075 Göttingen
Telefon: 0551-54724-0
Telefax: 0551-54724-21
www.cuvillier.de

1. Auflage, 2022
Gedruckt auf umweltfreundlichem, säurefreiem Papier aus nachhaltiger Forstwirtschaft.

ISBN 978-3-7369-7582-8
eISBN 978-3-7369-6582-9

Table of contents

1 Introduction and motivation

Over the last two centuries, economic growth has been driven at the expenses of the environment. In a scenario where the demand for energy constantly grows, fossil fuel-fired power generation still accounts for over 70% of the greenhouse gas emissions, caused mainly by the combustion of coal and oil. According to the International Energy Agency (iea), energy-related CO_2 emissions should be cut in half by 2050 compared to 2009 to limit the increase in the world average temperature by 2 °C. Despite the advances in renewable energy alternatives, their supply is strongly dependent on favorable weather conditions and their efficient storage is still difficult. Thus, the key factor for the transition towards a more sustainable future relies on decarbonizing power generation. Alone the substitution of heavy fossil fuels by natural gas is able to decrease the related emissions in approximately 50% [1–4]. Nevertheless, a developing interest has been shown for carbon-free fuel sources as ammonia and liquid hydrogen, which are potential alternatives to produce energy with a cleaner emission profile. However, the low heat of combustion and flammability of ammonia and the economically efficient storage and transport of liquid hydrogen still remain unsolved challenges for their sustainable utilization [5].

Besides the compactness, gas turbines can operate with versatile fuel sources and produce energy with high efficiency. For this reason, they have found increasing service in the past 60 years in the power industry both among utilities and industrial plants as well as for aviation throughout the world [6]. In combined cycle operation and with inlet temperatures exceeding 1400 °C, efficiencies as high as 63% can be achieved [2]. Therefore, different strategies are adopted to protect the currently used Ni-based superalloys such as the deposition of yttria-stabilized-zirconia thermal barrier coatings (TBC) and intensive film cooling. This standard is, however, not realistic when considering service for considerable amount of times ($t > 10{,}000$ h), since the great mismatch between the coefficient of thermal expansion (CTE) of the TBC and the alloy increase the risk of coating spallation and limit the application of metal-based components in turbine engines [7–10]. Especially envisioning the use of carbon-free fuel sources in future gas turbines, water vapor is one of the main products of combustion, which intensifies the degradation of these alloys [5,11–13]. Hence, advances towards reduced greenhouse gas emissions and more efficient gas turbines require their substitution by more robust materials resistant to oxidation and corrosion, able to endure service at higher temperatures.

Due to the reduced density, lower CTE (3 - 5.5 10^{-6} K^{-1}), high temperature creep resistance and melting points, non-oxide silicon-based ceramics as Si_3N_4, SiC and SiC/SiC composites stand out for application in combustion environments [14–21]. In oxidative environments, the formation of a thermally grown SiO_2-scale ensures a great oxidation stability. Nevertheless, $H_2O_{(v)}$ reacts with this scale at temperatures above 1200 °C, leading to corrosion and the rapid recession of the ceramic surface. Over the last three decades, progress has been made to develop reliable environmental barrier coatings (EBC) to hinder these detrimental effects [19,22–26]. Among the suitable EBC materials, rare-earth silicates have gained attention due to a high temperature endurance of at least 1400 °C and very low SiO_2 activity [10,19,27–29]. Despite the higher corrosion rates of the disilicates in comparison with monosilicates, the lower CTE results in a better thermomechanical compatibility to Si_3N_4, SiC and SiC/SiC substrates and should be considered when designing a reliable EBC for gas turbines. In special, ytterbium disilicate ($Yb_2Si_2O_7$, CTE = 3.91 10^{-6} K^{-1}) exhibits no polymorphic transition and is therefore an attractive EBC candidate [19,27,30,31].

Despite the suitable protection against corrosion, the diffusion of oxygen through the EBC during long-term service at high temperatures leads to the oxidation of the substrate and jeopardizes the mechanical stability of coated systems [32,33]. Therefore, thermomechanical compatible bond-coats based on mullite (CTE = 5.71 10^{-6} K^{-1}) or silicon (CTE = 4.1 10^{-6} K^{-1}) are applied [10,19,27,34–36]. In comparison with mullite, silicon bond-coats are dense and can effectively hinder the diffusion of oxygen by forming a slow-growing SiO_2-scale (CTE = 3.1 10^{-6} K^{-1}, β-cristobalite). Moreover, they have been reported to enhance adhesion, whereas oxide coatings usually exhibit an insufficient bonding strength to non-oxide silicon-based ceramics [10,27,37].

The most common techniques for the deposition of EBC systems onto non-oxide silicon-based ceramics are plasma spray [10,27,34,38], chemical (CVD) [39,40] or physical vapor deposition (PVD) [20,24]. PVD and CVD techniques are usually denoted by the very low deposition rates (1 μm h^{-1}), requiring extended processing times to achieve thick coatings [41]. The main disadvantages of plasma spraying lie within the evaporation of silica from the rare-earth silicate feedstock powder and in the rapid cooling rates during the coating deposition, resulting in an inhomogeneous phase composition and an amorphous microstructure. Due to the amorphous microstructure, plasma-sprayed coatings usually require a post-processing treatment at temperatures exceeding 1200 °C and annealing times of over 24 h in air to enable phase crystallization [27,34]. Besides the energy and time-consuming procedures, another main drawback of the mentioned techniques is the difficult coating of complex geometries, as required for gas turbines. In contrast, slurry-based techniques such as sol-gel or the polymer derived ceramics (PDC) enable the deposition of coatings by simple methods as spraying, spin- or dip-coating [42–46].

In special, the PDC route relies on the use of reactive silicon containing precursors as silazanes, carbosilanes and siloxanes. After or during the shaping process, a thermal treatment at low temperatures induces cross-linking, resulting in a thermoset, whereas further heating at above 400 °C yields a stable ceramic. Among the available precursors for coating application, silazanes stand out due to the commercial availability, high ceramic yield, strong adhesion to most substrates and high stability in oxidative and corrosive media [42,43,47]. Nevertheless, the protection of silazane-based coatings in such media often relies on the formation of a passivating SiO_2-scale, which undergoes detrimental reaction with water vapor in combustion environments. Moreover, mass loss and densification during the polymer to ceramic transition cause a volume shrinkage exceeding 50% [42,48], limiting the applicable thickness of crack and pore-free coatings to only a few micrometers [49,50]. The adverse effects related to the precursor shrinkage can be overcome by the use of fillers, hence enabling the deposition of thicker coatings with tailored properties [43,48,49,51–53].

These fillers are usually classified as passive, active and meltable. In contrast to passive fillers, active fillers undergo reaction with the atmosphere, precursor or pyrolysis products, thus expanding in volume and yielding tailored ceramic phases. The main role of meltable fillers is to act as densifying agents during melting, sealing porosity and relaxing stresses due to CTE mismatches at high temperatures. Consequently, the deposition of dense coatings of up to 100 μm for protection of steel substrates against oxidation and corrosion up to 700 °C was achieved [51]. Moreover in previous work [54], it was demonstrated that silazanes react with rare-earth oxides (RE_2O_3, RE = Y or Yb) as active fillers during pyrolysis in air to yield the respective silicates according to Eq. 1.1.1 and 1.1.2. Due to the shrinkage of the silazane precursor during pyrolysis, crack-free but porous coatings could only be obtained by limiting its amount, whereas the deficit of silicon resulted in unconverted rare-earth oxides.

$$RE_2O_3 + SiO_2 \rightarrow RE_2SiO_5 \qquad \text{(Eq. 1.1.1)}$$

$$RE_2O_3 + 2\ SiO_2 \rightarrow RE_2Si_2O_7 \qquad \text{(Eq. 1.1.2)}$$

A similar coating strategy was reported for the generation of ytterbium mono- and disilicate coatings on Si_3N_4 by the reaction of a polysiloxane and Yb_2O_3 during pyrolysis. The Si_3N_4 substrate was coated by dipping, followed by pyrolysis at 1000 °C in air and sintering at

1500 °C in N_2 atmosphere, achieving a coating thickness of 20 μm. However, no information about the formed crystalline phases was provided. Moreover, the coatings rapidly spalled-off during hot gas corrosion tests at 1450 °C for 200 h (p = 1 atm, p_{H2O} = 0.18 atm and v = 100 m s^{-1}) due the residual porosity and insufficient adhesion to the substrate [55].

1.1 Goals

As mentioned before, ytterbium disilicate ($Yb_2Si_2O_7$) is a very attractive material for the protection of non-oxide silicon-based ceramics in combustion environments due to the matching CTE, very low corrosion rate and no polymorphic crystalline transition. Plasma spraying, PVD and CVD are typical techniques for the deposition of EBCs, denoted by the energy and time-consuming procedures. By contrast, the PDC route stands out due to the low-cost and uncomplicated application of coatings by typical lacquer techniques based on reactive silicon-containing precursors. Despite the excellent adhesion and the proven potential of silazane-based coatings for protection of ceramic and metal substrates against oxidation and corrosion in harsh environments, the formation of SiO_2 limits their application in combustion environments. In this context, the conversion of forming SiO_2 into hot gas corrosion resistant phases like rare-earth silicates is imperative to increase the stability of silazane-based coatings.

As will be presented in Section 2.8, the high reactivity of silazanes with ytterbium oxide (Yb_2O_3) leads to the efficient formation ytterbium silicates upon pyrolysis in air. Nevertheless, monolithic samples based on this system are characterized by a high porosity content (≈ 40 vol%) [54]. In addition, stable but porous coatings can only be achieved by limiting the fraction of silazane, which results in considerable amounts of unconverted ytterbium monosilicate (Yb_2SiO_5) and Yb_2O_3 after pyrolysis. Due to the high CTE mismatch of these phases with non-oxide silicon-based ceramics, associated with the porous coating microstructure, only a limited protection against hot gas corrosion can be expected in combustion environments. Based on these premises and regarding the customary techniques for EBC application, the main goal of this work consisted in developing a straightforward and easier approach for the deposition of $Yb_2Si_2O_7$ coatings by the PDC route to hinder hot gas corrosion of non-oxide silicon-based ceramic components.

During the pyrolysis step of silazane-based coatings, their microstructure, composition and mechanical properties can be strongly influenced by the substrate. Therefore, the development of a slurry and the assessment of suitable pyrolysis parameters to yield dense and crystalline $Yb_2Si_2O_7$ should occur without the influence of a substrate by processing free-standing samples as powder and monoliths. Subsequently, coating experiments should be carried out with a non-oxide silicon-based ceramic substrate (i.e. Si_3N_4, SiC or SiC/SiC). In order to enable a straightforward and easy coating application, the largest coating thickness should be obtained with the lowest number of steps to reduce the processing time. In addition, the influence of the selected substrate on the resulting coating properties should be investigated.

During service in gas turbines, water vapor is a main product of combustion, which composes approximately 10% of the atmosphere. Corrosion by water vapor at above 1200 °C and a flow velocity that can exceed 100 m s^{-1} lead to the rapid volatilization of SiO_2 and the considerably high recession rates of ceramic components [11,15]. Besides hot gas corrosion, changes in the operating conditions of gas turbines such as starting or an emergency shut-down can cause harsh thermal shocks of more than 800 K within one second, leading to the evolution of additional thermal stresses [56], which must be endured by the coating system during the life expectancy of the components in gas turbines (t > 10,000 h). During the evaluation of the performance of the coating system in turbine-like environments, it should remain well-adhered and reduce hot gas corrosion regarding uncoated substrates.

Since gas turbines are expected to operate for considerable amount of times and rare-earth silicate-based environmental barrier coatings are slowly corroded in hot gas environments, the deposition of thick coatings (> 100 μm) is desired to extend the life span of coated ceramic parts. To obtain a coating system with high thickness, strong adhesion, low residual porosity and high thermomechanical stability are the greatest challenges of PDC processing, since most PDC-based coatings are limited to a few microns in thickness. Thus, alternatives should be sought to increase the achievable thickness of the PDC-based $Yb_2Si_2O_7$ coating. At last, coating experiments should be carried out with other non-oxide silicon-based ceramics, followed by the characterization of the resulting microstructure and composition to evaluate the transferability of the developed coating technology.

2 State of the art

2.1 Overview about gas turbines

The gas turbine has found increasing service in the past 60 years in the power industry both among utilities and industrial plants as well as for aviation throughout the world. Its compactness, low weight, and multiple fuel application make it an attractive candidate for energy production. To the day, different commercially available gas turbines run on natural gas, diesel fuel, methane and vaporized fuel oils. Depending on the availability, natural gas is often the fuel of choice because of its clean burning and competitive price. Its low carbon-to-hydrogen ratio means that it emits substantially less CO_2 than other fossil fuels, which can reduce emissions in up to 50% compared to coal powered energy plants and does not require a post-combustion waste treatment [2,4,6]. Nevertheless, cleaner energy sources as liquid hydrogen and ammonia have gained increasing interest and are expected to be extensively used as fuel alternatives in the near future [5].

Due to their high flexibility, gas turbines are available in different sizes and power outputs ranging from 20 kW to 600 MW, which are among the most efficient types of heat engines. In simple cycle configuration, gross efficiencies ranging from 30 to 46% are usually achieved. However, by using the heat from the gas turbine exhaust to produce steam in a heat-recovery steam generator in combined cycle operation, the production of additional electrical power can lead to overall efficiencies that exceed 60%. The two factors, which most affect turbine efficiencies, are pressure ratios and temperature. The increase in the pressure ratio increases the overall efficiency at a given temperature. However, beyond a certain point at any given firing temperature, this may result in lowering the cycle efficiency. In contrast, for every 56 °C increase in the firing temperature, the work output increases approximately 8-13%, which results in a simple cycle efficiency increase of 2-4% [2,4,6].

For unprotected metal components, long-term service at temperatures above 704 °C leads to hot corrosion by salt melts contained in the used fuels. Anyhow, different strategies are adopted to protect the currently used Ni-based superalloys such as the deposition of yttria-stabilized-zirconia thermal barrier coatings (TBC) and intensive film cooling to keep the material temperature below 1050 °C. Due to the combination of both protection mechanisms, gas turbines can operate with firing temperatures up to 1427 °C, achieving an overall efficiency of 66% in combined cycle operation [7–10]. However, during starting up an operation or an emergency shut-down, the components in gas turbines may be subjected to a temperature gradient of above 800 K within one second, which leads to a tensile stress of several hundreds of MPa and may cause the rapid spallation of the TBC due to the great mismatch between its coefficient of thermal expansion (CTE) with the alloy. These conditions can achieve an extreme point at the first stage of the gas turbine, where the gas flow velocity (265 m s^{-1}) and pressure (1.6 MPa) are high enough so that the heat transfer coefficient approaches that of a water quench [56,57].

The combustor, vanes and turbine blades constitute the most critical components in the hot section of existing high performance long-life gas turbines. The combustor and vanes operate at relatively high temperature and low stress conditions, whereas extreme conditions of stress and temperature can be expected to be faced by the turbine blades. In this case, the air flow velocity in combustors are generally limited to 42 m s^{-1} and the conditions of stress achieve about 35 MPa. In contrast, turbine blades are subjected to speed flows of over 100 m s^{-1} and stress conditions that can easily exceed 300 MPa [6,15,58]. Madhu [59] estimated the maximum stress of a gas turbine blade integrated disk of a jet engine achieved for different nickel-based superalloys (NI-90, MAR-M-247 and IN 718). Table 2.1.1 depicts a summary of the obtained results.

Table 2.1.1. Simulation of the maximum stress achieved for different Ni-based superalloys in jet engines, adapted from [59].

Material	**NI-90**	**MAR-M-247**	**IN 718**
RPM	29,000	29,000	29,000
Material temperature (°C)	725	725	725
Young's modulus at 725 °C (GPa)	153	171	157
Yield strength at 725 °C (MPa)	580	800	852
Ultimate tensile strength at 725 °C (MPa)	816	1002	933
Maximum achieved stress (MPa)	557	762	731

Besides these extreme conditions, thermal fatigue, oxidation, and corrosion also decrease the service life of the components in gas turbines. Especially the extreme conditions of stress, temperature and corrosion make the design of suitable materials for turbine blades a challenge. Materials characteristics in a turbine blade for high performance in the long-term include limited creep, high rupture strength, resistance to corrosion, good fatigue strength, low coefficient of thermal expansion and high thermal conductivity to reduce thermal strains [6,15,58]. Thus, due to the reduced density, low CTE (3 - 5.5 10^{-6} K^{-1}), high temperature creep resistance and melting points, non-oxide silicon-based ceramics as Si_3N_4, SiC and SiC/SiC composites stand out for application in combustion environments [14–21].

An increase in the operating temperature from 982 °C to 1371 °C by the use of uncooled ceramic blades provides an improvement in the fuel consumption of more than 20%, which implies that power almost doubles for the same engine size. In other words, the engine size could be reduced in half and still retain the same power output. Other advantages of ceramic materials are the good tolerance to fuel contaminants such as sodium and vanadium, which are present in low-cost fuels and are highly corrosive to currently used Ni-based superalloys. Besides that, ceramics are also over 50% lighter than the superalloys [6]. By the implementation of ceramic components in future gas turbines, operating temperatures that exceed 1500 °C are expected to be reached [10,56].

2.2 Non-oxide silicon-based ceramics

The substitution of Ni-based superalloys by ceramic components in gas turbines accounts for higher efficiencies and lower emissions, besides the reduction in the turbine overall weight and size as detailed in Section 2.1. For the long-term service at temperatures of 1200 °C and above, only non-oxide silicon-based ceramics like silicon nitride (Si_3N_4), silicon carbide (SiC) and fiber reinforced silicon carbide matrix composites (SiC/SiC) are suitable [14–21]. In comparison with SiC, Si_3N_4 stands out due to the higher fracture toughness caused by the formation of needle-like β-Si_3N_4 crystals from α-Si_3N_4 during sintering, capable of deflecting cracks. However, due to the brittle behavior of monolithic ceramics, their application is rather meaningful as structural components for stationary gas turbines. In non-stationary turbines as aircraft engines, additional problems may occur such as the foreign object damage (FOD) or strong vibration, which require an even higher fracture toughness and damage tolerance. For this purpose, only non-oxide ceramic matrix composites (CMCs) as SiC/SiC are suitable [60–62]. In this section, a summary of the processing routes and resulting properties of the different Si_3N_4 and SiC ceramics and SiC/SiC composite will be given.

2.2.1 *Silicon nitride (Si_3N_4)*

In comparison with other high-performance ceramics, Si_3N_4 stands out due to the combination of good mechanical properties at room and high temperatures. Owing to the strong covalent nature of the Si-N bonds of Si_3N_4, its sintering through a solid-state reaction is practically impossible. At ambient pressure, this ceramic has no melting point, however it starts decomposing at a temperature of 1877 °C. Two processing routes are usually adopted for the processing of Si_3N_4 ceramics. The first one consists of nitriding elemental Si, hence leading to the formation of Si_3N_4, which is named as reaction bonded silicon nitride (RBSN). The advantage of this method is the low volumetric shrinkage of less than 1%. Anyhow, a residual porosity of 10 to 30 vol% is usually obtained, which is associated with a low fracture toughness of less than 400 MPa. The second method and more efficient to obtain dense ceramic bodies consists of mixing 1 μm Si_3N_4 powders with additives, followed by sintering, commonly classified as pressureless sintering (SSN, up to 0.1 MPa), gas pressure sintering (GPSN, up to 10 MPa), hot isostatic pressing (HIPSN, up to 200 MPa) and hot-pressing (HPSN).

The most commonly used sintering additives consist of binary oxides like Y_2O_3, rare-earth oxides, Al_2O_3 and MgO. During heating of the processed green body, the reaction of these oxides with SiO_2 available at the surface of the Si_3N_4 feedstock powder forms silicate melts, which densify the microstructure of Si_3N_4. This causes the solubilisation of the α-Si_3N_4 phase and its precipitation as β-Si_3N_4. Due to the hexagonal crystalline structure, the β-Si_3N_4 phase is known by its needle-like crystals and is often referred to as an *in situ* reinforcement. After sintering, most sintering additive systems form an amorphous or semi-crystalline grain boundary phase. The mechanical properties of Si_3N_4 are usually dependent on the raw materials, sintering additive content and their composition, as well as the processing route. The density varies according to the employed sintering additive system, usually lying between 3.2 and 3.6 g cm^{-3}. In addition, a good thermal shock resistance is reported due to the low CTEs of approximately 3 to 3.2 10^{-6} K^{-1} and the presence of the β-Si_3N_4 phase. The most important mechanical properties of the different Si_3N_4 ceramics according to the processing route are listed in Table 2.2.1 [61].

Table 2.2.1. Common properties of Si_3N_4 materials according to the processing route [61].

Type	Young's modulus (GPa)	Flexural strength (RT, MPa)	Flexural strength (1400 °C, MPa)	Fracture toughness (MPa $m^{1/2}$)
SSN	310 - 320	600 - 1000	100 - 300	5 - 7
GPSN	310 - 320	800 - 1300	100 - 300	5 - 10
HIPSN	310 - 320	800 - 1500	200 - 400	5 - 7
HPSN	310 - 320	1000 - 1700	200 - 400	5 - 7
RBSN	100 - 150	200 - 400	200 - 400	2 - 3

Despite the good mechanical properties at room temperature, a notorious decrease in the flexural strength at high temperatures is directly correlated to the softening of the amorphous grain boundary phase. Nevertheless, the crystallization of the grain boundary phase with an additional thermal handling in air at above 1000 °C improves the high temperature properties due to the reduced amount of low viscosity phases [63].

2.2.2 Silicon carbide (SiC)

Besides Si_3N_4, silicon carbide also stands out for high temperature application. Attractive properties of this ceramic are for example the high hardness, high temperature flexural strength and high thermal conductivity. Similarly to Si_3N_4, the sintering of SiC is extremely difficult and usually requires temperatures above 1900 °C, depending on the selected processing route. Some common processing routes are for example the pressureless sintering (SiC), liquid phase sintering (LPSiC), hot-pressing (HPSiC), hot isostatic pressing (HIPSiC) and silicon infiltration (SiSiC). Besides SiSiC, the sintering of SiC is achieved through the use of Al, B, C and their respective compounds like B_4C, BN, AlB_4 or SiB_6 as sintering additives. In comparison with Si_3N_4, the higher decomposition temperature of SiC at above 2500 °C enables its sintering with a lower additive content ranging from approximately 0.2 to 3 wt% [61,64,65].

At temperatures above 2100 °C, the low temperature phase β-SiC transforms into α-SiC, which is accompanied by a grain growth, desired for *in situ* reinforcement properties. In comparison with the other sintering techniques, silicon infiltration enables the sintering of SiC at relatively low temperatures due to the melting and infiltration of the microstructure by Si or Si-based alloys at around 1414 °C, also limiting its application to this temperature due to the residual amount of Si. Depending on the sintering method, different densities (3.12 to 3.21 g cm^{-3}) and CTE$\alpha_{30/1500}$ (3.60 - 4.9 10^{-6} K^{-1}) are reported as well as different grain boundary compositions, which influence directly the mechanical properties of SiC [61]. Table 2.2.2 summarizes the mechanical properties of SiC ceramics obtained by some of the common sintering methods.

Table 2.2.2. Common properties of SiC materials according to the processing route [61].

Type	**Young's modulus (GPa)**	**Flexural strength (RT, MPa)**	**Flexural strength (1400 °C, MPa)**	**Fracture toughness (MPa $m^{1/2}$)**
SiSiC	400	350 - 370	200	3 - 4.7
SSiC	410	430	450	2.5 - 4.7

LPSiC	420	600 - 800	400	4 - 6
HPSiC	440	500 - 650	500 - 650	3 - 4
HIPSiC	450	650	600	3 - 4

2.2.3 Fiber reinforced silicon carbide matrix composites (SiC/SiC)

In non-stationary gas turbines, additional problems like foreign object damage (FOD) or strong vibration may occur, whereas the brittle behavior of monolithic ceramics as Si_3N_4 and SiC can lead to a sudden failure of the ceramic components. The inclusion of a second phase in monolithic matrices, as for example high tensile strength fibers, can increase their damage tolerance and fracture toughness. The weak bonding between matrix and fiber is crucial to dissipate energy during crack formation. Different mechanisms contribute to stopping crack propagation, like fiber pull-out, crack deflection and branching [60].

For application at temperatures above 1100 °C, only non-oxide ceramic matrix composites (CMCs) are suitable [19]. Different classes of SiC CMCs are commercially available as SiC/SiC, C/C-SiC and C/SiC. C/C-SiC and C/SiC are commonly processed by liquid phase infiltration of C/C composites or C/SiC with Si at temperatures above 1400 °C [66]. However, the oxidation of carbon at temperatures above 450 °C leads to a strong decay in their mechanical properties, hindering the application in oxidative atmospheres [60,62,67,68]. Especially for use in gas turbines, where temperatures can exceed 1400 °C, only SiC/SiC ceramic composites are suitable. Among the various fabrication methods of SiC/SiC CMCs, chemical vapor infiltration (CVI) is preferred due to the high purity of the deposited SiC matrix, hence achieving a thermal stability up to 1400 °C without considerable loss of the mechanical properties [60,69]. Another attractive property of CVI SiC/SiC is the strength retention after thermal shock cycles involving heating up to the desired temperature and quenching in water at 20 °C [69].

To achieve a good performance during long-time exposure at high temperatures, it is important that the fibers exhibit high strength, high stiffness, low density and high thermal and chemical stability. Therefore, thin SiC fibers (∅ 8-15 μm) with a tensile strength of 2-3.5 GPa are used for the processing of SiC/SiC composites [60]. The weak bonding between fiber and matrix is assured by the deposition of coatings with thicknesses of approximately 0.1-0.5 μm usually based on pyrolytic carbon, boron nitride or silicon-doped boron nitride. Nevertheless, for high environmental durability and high compliance for matrix crack deflection, the interphase and composition of the coatings is typically based on silicon-doped boron nitride [62]. Afterwards, the SiC matrix is applied by CVI.

CVI is a slow process, and the obtained composite materials possess some residual porosity and density gradients. Despite these drawbacks, the CVI process presents a few advantages: (i) the strength of reinforcing fibers is not affected during composite manufacture, (ii) the nature of the deposited material can be changed easily, simply by introducing the appropriate gaseous precursors into the infiltration chamber, (iii) a large number of components, and (iv) large complex shapes can be produced in a near net shape. In the final steps, SiC particulate is slurry cast into the component preform near room temperatures, followed by a finishing step in which molten silicon is infiltrated at approximately 1400 °C into any open remaining matrix porosity [62,69]. An alternative processing route for the densification of CVI-SiC/SiC is carried out by the infiltration of the matrix with an inorganic Si-based polymer followed by pyrolysis. An additional infiltration with molten silicon forms *in situ* SiC upon reaction with residual carbon. Both processing routes always result in some residual silicon in the matrix, allowing these CMCs to display undesirable matrix creep that begins at 1100 °C and becomes excessive above 1300 °C [60,62]. Moreover, certain properties of the resulting SiC/SiC composites may be influenced by various factors as the reinforcing fiber architecture, the used SiC fibers, matrix properties and the fiber/matrix bonding strength. Table 2.2.3 shows general properties of CVI-SiC/SiC composites reinforced with 0/90 balanced Nicalon™ fabrics at room temperature (RT) as well as at 1000 and 1400 °C [69].

Table 2.2.3. Mechanical and thermophysical properties of CVI-SiC/SiC composites reinforced with 0/90 balanced Nicalon™ fabrics [69].

	Temperature		
Property	**RT**	**1000 °C**	**1400 °C**
Fiber content (vol%)	40	40	40
Density (g cm^{-3})	2.5	2.5	2.5
Porosity (vol%)	10	10	10
Strain-to-failure	200	200	150
Young's modulus (GPa)	230	200	170
Flexural strength (MPa)	300	400	280
Fracture toughness (MPa m$^{1/2}$)	30	30	30
In-plane compressive strength (MPa)	580	480	300
Through-thickness compressive strength (MPa)	420	380	250

In-plane CTE (10^{-6} K^{-1})	3	3
Through-thickness CTE (10^{-6} K^{-1})	1.7	3.4

2.3 Oxidation and hot gas corrosion behavior of Si_3N_4

In oxidative environments, the formation of a thermally grown SiO_2-scale at the surface of Si_3N_4, SiC and SiC/SiC ceramic components ensures a great oxidation stability according to the mechanism of Eq. 2.3.1 (see example for Si_3N_4). Nevertheless, water vapor ($H_2O_{(g)}$) is a product of combustion, accounting for at least 10 vol% of the emission gases, which also leads to oxidation (Eq. 2.3.2) [15,16]. At temperatures above 1200 °C and ambient pressure, the reaction of water vapor with forming SiO_2 leads to the significant volatilization of silicon hydroxide, according to Eq. 2.3.3. This leads to corrosion and the rapid recession of the ceramic structural components, which is intensified at higher temperatures or higher water vapor partial pressures (p_{H2O}) [11]. Therefore, progress has been made to develop reliable environmental barrier coatings (EBC) to hinder these detrimental effects [14,19,22–26].

$$Si_3N_4 + 3\ O_2 \rightarrow 3\ SiO_2 + 2\ N_{2(g)} \qquad \text{(Eq. 2.3.1)}$$

$$Si_3N_4 + 6\ H_2O_{(g)} \rightarrow 3\ SiO_2 + 4\ NH_{3(g)} \qquad \text{(Eq. 2.3.2)}$$

$$SiO_2 + 2\ H_2O_{(g)} \rightarrow Si(OH)_{4(g)} \qquad \text{(Eq. 2.3.3)}$$

In the past, different field tests were carried out with ceramic structural components for the hot section of gas turbines [14]. Solar Turbines initiated a 100 h engine test with a SiC/SiC CMC combustor liner and Si_3N_4 vanes and blades in order to improve the engine performance and to reduce the exhaust emissions. The engine was subjected to cold and hot restart and shut-down cycles that progressively increased in severity. After 68 h, the test was terminated due to the fracture of the Si_3N_4 vanes associated to the loss of the normally protective silica scale and dynamic fatigue of the material.

Similar tests were carried out by Rolls-Royce with Si_3N_4 vanes, which were submitted to average temperature and pressure of 1066 °C and 8.7 atm, respectively, whereas temperatures as high as 1288 °C were measured in the combustor. The inlet gas velocity at the vane section was about 162 m s^{-1} and the water vapor fraction for the gas entering the vanes was calculated to be 10% due to the extremely humid conditions during testing. After 815 h, no failures occurred. However, dimensional measurements indicated that the Si_3N_4 vanes exhibited a significant material recession.

Several other tests with ceramic components in gas turbine engines have been carried out by different enterprises [70], whereas oxidation, corrosion of the forming SiO_2 and surface recession were among the most critical problems. Since Si_3N_4 was used as a substrate for the development of the $Yb_2Si_2O_7$ coating system in this work, the oxidation and hot gas corrosion behavior of this ceramic material will be described in detail.

2.3.1 Dry environments

In general, the oxidation of silica formers such as Si, Si_3N_4 and SiC follows a parabolic mass gain at constant temperatures, where the diffusion of oxygen from the atmosphere through the forming SiO_2 passivating layer controls the oxidation rate [71,72]. A parabolic mass gain is also reported for sintered Si_3N_4 ceramics containing additives, however the oxidation rate and the high temperature properties are extremely dependent on the sintering additive system, its content and the level of impurities present in the Si_3N_4 feedstock powder.

Kiehle *et al.* [73] investigated the oxidation behavior of hot-pressed Si_3N_4 ceramics from 600 to 1450 °C and annealing times varying from 15 min to 96 h in air. Emission spectroscopy determined that the major impurities in the Si_3N_4 samples before oxidation corresponded to 0.3 Fe, 0.18 Al, 0.6 Mg, 0.1 Mn and 0.08 Ca (in wt%). Already at 750 °C for 24 h, the authors determined the formation of an amorphous SiO_2-scale. As the heating time or temperature increased, the formation of cristobalite crystals was reported, favored by the presence of small amounts of cation impurities. At above 1000 °C, MgO and CaO impurities in the Si_3N_4 samples reacted to form silicate products as $MgSiO_3$, $Ca_2MgSi_2O_7$, Mg_2SiO_4 and $CaMg(SiO_3)_2$. The authors concluded that although the level of impurities of Ca and Mg in the bulk Si_3N_4 corresponded to only 0.6 and 0.08 wt%, respectively, their continuous uptake led to the detection of up to 12 wt% within the formed oxidation layer.

The oxidation behavior of hot-pressed Si_3N_4 ceramics containing, respectively, MgO and Y_2O_3 was investigated by Cubiciotti and Lau in the temperature range from 1248 to 1523 °C up to 25 h in dry oxygen [74,75]. For both hot-pressed materials, the authors reported a parabolic mass gain, which is expected for SiO_2 formers. However, they concluded that the outward diffusion of the additive cations from the bulk of Si_3N_4 controlled the oxidation rate of these materials.

Klemm *et al.* [76] investigated the long-term stability of Si_3N_4 ceramics processed by hot-pressing without sintering additives and with varying contents of Y_2O_3, Al_2O_3 and $MoSi_2$ as sintering additives at 1500 °C for up to 10,000 h in air. In comparison with pure Si_3N_4, oxidation rates of at least an order of magnitude were reported for sintering additive containing Si_3N_4 ceramics. Among the investigated materials (SNY (10 wt% Y_2O_3), SNYAl (10 wt% Y_2O_3 and 0.6 wt% Al_2O_3) and SNYMo (10 wt% Y_2O_3 and 10 wt% $MoSi_2$)), the highest oxidation rate was reported for SNYAl, explained by the faster diffusion of Y^{3+} and Al^{3+} cations from sintering additives into the oxidation layer, which favored the change of its chemistry. This led to a reduction of its viscosity and the formation of low temperature eutectic phases, enabling the easier diffusion of oxygen and the oxidation of the underlying Si_3N_4. The best oxidation performance was attributed to the SNYMo material, which was able to retain 65% of its flexural strength (500 MPa) after 10,000 h of oxidation.

In summary, the use of a pure Si_3N_4 feedstock is essential for gas turbine application, since low amounts of impurities as much as 80 ppm of Ca^{2+} and F^- in the grain boundary phase can lead to a strong decay of the flexural strength of Si_3N_4 ceramics at temperatures higher than 1100 °C. Moreover, the standard sintering additive systems containing MgO and Al_2O_3 are less beneficial for the development of high-temperature Si_3N_4 due to the formation of eutectic phases at low temperatures, increasing the oxidation rates. The addition of Y_2O_3 and other rare-earth oxides shifts the formation of eutectic phases to higher temperatures, whereas the more viscous SiO_2 passivating layer protects the underlying Si_3N_4 ceramics more effectively [19].

Several studies have been performed in order to replace the standard $Y_2O_3/Al_2O_3/MgO$ sintering additive system by other oxide systems forming high refractory silicates as Sc_2O_3 or the rare-earth oxides from the lanthanide group [17,19,77,78]. The sintering of these materials is usually more difficult in comparison with the standard sintering additive system due to the higher temperatures necessary to generate glassy grain boundary phases. To the day, two commercially available turbine grade Si_3N_4 ceramics based on lanthanide sintering additive systems with low level of impurities are known, namely AS800 from Honeywell Aerospace (USA) and SN282 from Kyocera Corporation (Japan). Chemical analysis performed by inductively coupled plasma mass spectrometry (ICP-MS) with both materials determined 4.2 wt% La, 1.3 wt% Y and 1.0 wt% Sr for AS800 and 7 wt% Lu for SN282, indicating sintering additives from the systems $La_2O_3/Y_2O_3/SrO$ and Lu_2O_3, respectively [17].

In comparison with other standard Si_3N_4 materials, Si_3N_4 ceramics containing rare-earth based sintering additive systems exhibited considerably lower oxidation rates after oxidation at 1500 °C for 1000 h in air and an enhanced creep resistance at 1400 °C [19,76]. The high oxidation resistance of these materials was attributed to the highly viscous amorphous grain-boundary phase and to the lower diffusion rates of large rare-earth cations (RE^{3+}) in comparison with additives from the system $Y_2O_3/Al_2O_3/MgO$ [78]. Besides the lower oxidation rates, the use of rare-earth oxides as sintering additives is also interesting for application in wet combustion environments due to the formation of corrosion resistant phases as rare-earth silicates upon oxidation, which may contribute to reduce the volatilization of SiO_2 from the oxidized Si_3N_4 ceramic [17,78].

2.3.2 *Wet environments*

In combustion environments, water vapor acts as the primary oxidant species regardless of the oxygen partial pressure due to the higher permeability of H_2O through the forming SiO_2-scale, accounting for an oxidation rate 10-20 times higher than in dry environments [15,71]. As mentioned before, water vapor still causes the volatilization of SiO_2 as $Si(OH)_4$ [11,16], that has been overseen by most of the works dealing with the oxidation of Si_3N_4 ceramics in wet environments until recently.

Singhal [79] investigated the effect of water vapor on the oxidation of MgO containing hot-pressed Si_3N_4 in dry oxygen and in wet oxygen containing 3.3% water vapor over the temperature range of 1200 to 1400 °C. In comparison with dry environments, higher oxidation rates were measured for wet environments, attributed to the diffusion of OH^- ions through the surface silica and to the diffusion of the additive and impurity cations, reducing its viscosity. Opila and Jacobson [18] described that the viscosity of the SiO_2-scale can also be reduced by the transport of impurities as Na and K from the environment in form of stable gaseous hydroxides by water vapor, contaminating it and enabling the higher permeation of oxidative species.

Different hot-pressed Si_3N_4 ceramics mostly containing sintering additives from the system $Y_2O_3/Al_2O_3/MgO$ were oxidized by Maeda *et al.* at 1300 °C for up to 100 h in wet air ($p = 1$ atm, $p_{H2O} = 0$ to 0.4 atm) [80,81]. The authors described a linear relationship between the measured mass gain and the p_{H2O}. The water vapor in the atmosphere strongly accelerated the devitrification of amorphous silica into cristobalite. Moreover, no effect on the flexural strength of the investigated samples was determined at this temperature regardless of the water vapor partial pressure.

Fox *et al.* [16] investigated the oxidation behavior of CVD-Si_3N_4 in comparison with the Si_3N_4 ceramics containing rare-earth-based sintering additives AS800 and SN282. Samples with the dimensions of approximately 25x7x3 mm were exposed to dry and wet oxygen atmospheres (50% H_2O) in thermogravimetry experiments with a flowrate of 4.4 cm s^{-1} and temperatures between 1200 and 1400 °C. In general, higher oxidation rates were measured for the samples exposed to wet environments. In contrast to AS800 and SN282, the formed SiO_2-scale at the surface of the CVD-Si_3N_4 sample was amorphous at 1200 °C, which only crystallized at 1300 °C. Moreover, the exposure of the SN282 material to temperatures above 1200 °C led to the additional crystallization of $Lu_2Si_2O_7$. In both dry and wet oxygen atmospheres, the specific weight change of CVD-Si_3N_4 was lower than the sintering additive containing samples. For dry oxidation, a parabolic mass gain was reported for all the samples, whereas in wet atmospheres, corrosion and volatilization of SiO_2 by water vapor led to mass loss and surface recession.

Klemm [20] also investigated the recession behavior of AS800 and SN282 silicon nitride materials in comparison with hot-pressed Si_3N_4 ceramics containing Mo, Y_2O_3 and Yb_2O_3 as sintering additives in burner rig tests at 1400 °C for up to 100 h (p = 5 atm, p_{H2O} = 0.5 atm, v = 50 m s^{-1}). Thereby, a corrosion rate of approximately 0.25 mg cm^{-2} h^{-1} was measured for the hot-pressed specimens and SN282. Unlike expected, a corrosion rate of 0.57 mg cm^{-2} h^{-1} was determined for AS800, which was attributed to the used sintering additive system composed by Y_2O_3, La_2O_3 and SrO and its higher content in comparison with the other tested materials. The high content of Y, La and Sr cation impurities in the forming SiO_2-scale reduced its viscosity, which intensified oxidation and led to higher recession rates.

As a complement to the work of Fox *et al.* [16], Opila *et al.* [17] investigated the oxidation and recession behavior of AS800 and SN282 silicon nitride materials in high-pressure burner rig (HPBR, p = 6 atm, p_{H2O} = 0.6 atm, v = 20 m s^{-1}, T ranging from 1150 to 1330 °C for up to 196 h) and in a turbine engine. In the latter case, a Rolls-Royce Allison 501-K turbine was retrofitted with AS800 and SN282 ceramic vanes in a natural gas processing plant operated by ExxonMobil near Mobile, AL, USA. The average operating temperature corresponded to 1066 °C with 8.7 atm pressure, p_{H2O} = 0.87 atm, and inlet gas velocities at the vane mid-span of about 161 m s^{-1}, accelerating up to 575 m s^{-1} at the vane exit. The AS800 vanes were exposed in the test for 815 h and the SN282 for 1818 h. After the tests, the surface of the Si_3N_4 materials was enriched with rare-earth silicates, which possess a high stability in combustion environments. Nevertheless, the formation of rare-earth disilicate phases was not continuous, which offered little additional protection against the volatilization of silica, resulting in the linear recession of both materials. The authors concluded that the protection of sintering additive containing Si_3N_4 by environmental barrier coatings is imperative for application in combustion environments.

2.4 Suitable environmental barrier coating materials

In order to protect non-oxide silicon-based ceramics against hot gas corrosion and surface recession in combustion environments, different materials were investigated as suitable candidates for application as environmental barrier coatings. The recession behavior of these materials has been investigated in wet environments and in temperatures ranging from 1000 to 1500 °C, simulating the harsh conditions in gas turbines.

Fritsch *et al.* [23] investigated the recession behavior of Al_2O_3, ZrO_2, mullite, $ZrSiO_4$ and yttrium aluminium garnet (YAG) in a burner rig at temperatures between 1200 and 1500 °C, p = 1 atm, p_{H2O} = 0.24 atm and gas flow velocity of 100 m s^{-1} for up to 300 h. In the tested temperature range, ZrO_2 showed absolutely no corrosion, whereas the other materials suffered degradation at temperatures above 1300 °C, due to the volatilization of $Si(OH)_4$ and $Al(OH)_3$. Among the tested materials, Al_2O_3 and mullite showed the highest corrosion rates, corresponding to approximately 4.8 10^{-2} mg cm^{-2} h^{-1} at 1500 °C.

The stability of different rare-earth oxides (RE_2O_3, RE = Sc, Dy, Er, Yb and Y) was evaluated by Courcot *et al.* [82] at temperatures ranging from 1000 to 1400 °C in wet air (p = 1 atm and p_{H2O} = 0.5 atm) under flowing gas velocity of 5 cm s^{-1}. The authors determined the lowest corrosion rate for Sc_2O_3 at 1400 °C (7.4 10^{-2} mg cm^{-2} h^{-1}), whereas a corrosion rate of approximately 13 10^{-2} mg cm^{-2} h^{-1} was measured for the other oxides, caused by the volatilization of REOOH and $RE(OH)_3$ species.

The corrosion behavior of $ASiO_4$ (A= Ti, Zr and Hf) was examined at 1500 °C for up to 50 h by Ueno *et al.* in static water vapor environment (p = 1 atm and p_{H2O} = 0.3 atm) [24]. The authors measured a corrosion rate of 7.08 10^{-3} and 1.33 10^{-3} mg cm^{-2} h^{-1} for $HfSiO_4$ and $ZrSiO_4$, respectively, whereas the mass of $TiSiO_4$ remained unchanged. The mass loss was attributed to the reaction of the silicates with water vapor, leading to the volatilization of $Si(OH)_4$ and the remaining of AO_2 species. Despite the highest corrosion rate, $HfSiO_4$ was recommended as an EBC material due to the matching coefficient of thermal expansion to non-oxide Si-based ceramics.

A study conducted by Lee *et al.* [22] evaluated the hot gas corrosion stability of different rare-earth monosilicates (RE_2SiO_5, RE = Y, Er, Yb and Lu) in comparison with rare-earth disilicates ($RE_2Si_2O_7$, RE = Sc and Yb) and the corrosion resistant glass-ceramics BAS (BaO-Al_2O_3-$2SiO_2$), SAS (SrO-Al_2O_3-$2SiO_2$) and BSAS (1-xBaO-xSrO-Al_2O_3-$2SiO_2$, $0 \leq x \leq 1$). Dense samples with dimensions of 25x12.5x1.5 mm were prepared with the corresponding materials and exposed to water vapor containing atmosphere (p = 1 atm, p_{H2O} = 0.5 atm, v = 4.4 cm s^{-1}) at 1500 °C for up to 100 h in thermogravimetric experiments. The hot gas corrosion stability of the investigated materials followed the order Yb_2SiO_5 > Er_2SiO_5 > Y_2SiO_5 > Lu_2SiO_5 > $Yb_2Si_2O_7$ > $Sc_2Si_2O_7$ > BSAS > SAS > BAS. The authors concluded that the recession of the corrosion resistant glass-ceramics BAS, SAS and BSAS to be unacceptably high at this temperature for long-term application as EBCs in advanced gas turbines.

Ridley and Opila [31] investigated the kinetics of corrosion and the phase stability of $Yb_2Si_2O_7$ sintered bodies in high-velocity water vapor in the temperature range from 1200 to 1400 °C for up to 250 h. During the hot gas corrosion tests, deionized water was pumped into a 1 mm diameter Pt-Rh capillary passing through a preheater at 900 °C and injected into the hot zone of the tube furnace, resulting in gas flow velocities of up to 262 m s^{-1}. The authors detected Yb_2SiO_5 after corrosion at 1200 °C due to the volatilization of $Si(OH)_4$ from $Yb_2Si_2O_7$, resulting in a porous microstructure. At above 1300 °C, the further corrosion of the Yb_2SiO_5 phase led to the generation of Yb_2O_3 and an increased porosity content. Corrosion rates of 0.8 10^{-2}, 2.2 10^{-2} and 4.2 10^{-2} mg cm^{-2} h^{-1} were measured at 1200, 1300 and 1400 °C, respectively, and the rate-controlling mechanism was attributed to the formation and outward diffusion of $Si(OH)_4$ from the reaction interfaces.

Klemm [19] published the corrosion rates of various oxide and non-oxide ceramic materials in hot gas environments according to the results of the Ph.D. thesis of Marco Fritsch, FhG IKTS Dresden [83]. Bending bars with the dimensions of 3.8x3.0x36 mm were processed with the ceramic materials, which were corroded in an atmospheric burner rig at 1450 °C for up to 700 h (p = 1 atm, p_{H2O} = 0.28 atm, v = 100 m s^{-1}). Table 2.4.1 exhibits the measured corrosion rates and the corresponding coefficients of thermal expansion (CTE). Despite the stability of Yb_2SiO_5, $Y_2Si_2O_5$ and ZrO_2 in the tested environments, the generation of stresses can be expected between these phases and SiC and Si_3N_4 ceramics due to the great CTE mismatch. Hence, $Y_2Si_2O_7$ and $Yb_2Si_2O_7$ were recommended as suitable EBC materials due to the matching CTEs and very low corrosion rates.

Table 2.4.1. Summary of the corrosion rates in atmospheric burner rig (1450 °C, p = 1 atm, p_{H2O} = 0.28 atm, v = 100 m s^{-1}) and the respective CTEs, adapted from [19].

Material	Corrosion rate (10^{-2} mg cm^{-2} h^{-1})	$CTE_{RT-1450\,°C}$ (10^{-6} K^{-1})
Si_3N_4	5.2	3.60
SiC	4.6	3.60
Al_2O_3	4.0	9.02
Mullite	3.8	5.71
$MoSi_2$	3.0	9.48
$ZrSiO_4$	1.5	6.96
Sc_2SiO_5	0.6	6.65
$Sc_2Si_2O_7$	0.5	5.37
$Y_2Si_2O_7$	0.3	4.07
$Y_3Al_5O_{12}$ (YAG)	0.2	9.18
$Yb_2Si_2O_7$	0.2	3.91
Y_2SiO_5	stable	6.67
Yb_2SiO_5	stable	6.65
ZrO_2 (3 mol% Y_2O_3)	stable	12.66
Y_2O_3	small weight gain	8.7*

* Measured from 25 to 1400 °C [84].

2.4.1 *Summary*

In face of the results presented in this section, the candidates ZrO_2, Y_2SiO_5 and Yb_2SiO_5 stand out as EBC materials, as they resist corrosion even after 100 h at 1450 °C (p = 1 atm, p_{H2O} = 0.28 atm and v = 100 m s^{-1}), followed by $Yb_2Si_2O_7$, YAG and $Y_2Si_2O_7$. The highest corrosion rates were reported for mullite, Al_2O_3 and the rare-earth oxides, which are comparable to those of non-oxide silicon-based ceramics. However, besides $Yb_2Si_2O_7$ and $Y_2Si_2O_7$, the large CTE mismatch between the other tested EBC candidates with Si_3N_4 and SiC would likely result in a low thermomechanical stability, hence leading to the extensive formation of cracks and the spallation of the deposited environmental barrier coatings during thermal cycling. In contrast, the disilicate phases possess low CTEs (4.07 10^{-6} K^{-1} for $Y_2Si_2O_7$, and 3.91 10^{-6} K^{-1} for $Yb_2Si_2O_7$), resulting in a small mismatch with those of Si_3N_4 and SiC (CTE ≈ 3.60 10^{-6} K^{-1}), besides the very low corrosion rates. The main disadvantage of $Y_2Si_2O_7$ is the excessive amount of reversible polymorphic transformations, which take place at 1225 °C ($\alpha \rightarrow \beta$) and 1445 °C ($\beta \rightarrow \gamma$) [85]. Materials having polymorphs with different densities are not desirable as coating candidates, since the phase transformation is accompanied by a volume change according to the operating temperature, which lead to the evolution of stresses in the coating system. In contrast, $Yb_2Si_2O_7$ presents no polymorphism [30] and is therefore a suitable candidate as environmental barrier coating for non-oxide silicon-based ceramics.

2.5 Environmental barrier coatings for Si_3N_4, SiC and SiC/SiC

Major improvements in turbine inlet temperatures can be achieved by the use of non-oxide silicon-based ceramic components. Nevertheless, the lack of environmental durability of these ceramics in combustion environments is explained by the reaction of water vapor with SiO_2. Therefore, the development of suitable environmental barrier coating (EBC) systems to hinder these detrimental effects has been a topic of great interest over the last 30 years.

2.5.1 Coatings of the 1st, 2nd and 3rd generation

Initially, mullite attracted great attention due to the low coefficient of thermal expansion (CTE), chemical compatibility with Si-based ceramics and good adherence. The first environmental barrier coatings were processed by air plasma spray, which resulted in a residual amount of amorphous mullite within its microstructure. The subsequent exposure of coated systems at temperatures above 1000 °C led to the crystallization of the amorphous phase, which is accompanied by shrinkage and caused the cracking and delamination of the coatings [22]. Another major issue with the mullite coating was the low stability in hot gas environments, whose corrosion rate is comparable to that of SiC and Si_3N_4 [19]. The volatilization of SiO_2 from the mullite coating left a porous Al_2O_3 surface which readily spalled off, caused by the great CTE mismatch of the oxide with non-oxide Si-based ceramics.

2.5.1.1 1st generation

In the first generation of EBCs, the problem with volatilization of mullite was overcome by the deposition of an yttria-stabilized-ZrO_2 (YSZ) top-coat, which was already in use as a thermal barrier coating for Ni-based superalloys in gas turbines and is resistant to corrosion by water vapor. This coating system could protect SiC ceramics for a few hundred hours at 1300 °C. However, the large CTE mismatch to the mullite bond-coat and the SiC substrate caused severe cracking and delamination, leading to the premature failure of the EBC system.

2.5.1.2 2nd generation

An EBC system with increased durability was developed within the NASA High Speed Research-Enabling Propulsion Materials (HSR-EPM) Program in joint research between NASA, General Electric and Pratt & Whitney [86,87]. Due to the lower CTE and a better thermomechanical compatibility to the mullite bond-coat and non-oxide Si-based ceramics, the

YSZ coating was replaced by BSAS ($1\text{-}xBaO\text{-}xSrO\text{-}Al_2O_3\text{-}2SiO_2$, $0 \leq x \leq 1$) [88]. The EBC performance was further enhanced by the introduction of a Si bond-coat, which provided a significantly better adherence. Hence, the coatings of the second generation consisted of a Si bond-coat, a mullite or a mullite + BSAS intermediate coating and a BSAS top-coat.

This coating system was scaled up and applied onto SiC/SiC CMC combustors liners used in three Solar Turbine Centaur 50s gas turbine engines [89,90]. By the end of 1999, the combustor had resisted over 24,000 h without failure at a maximum firing temperature of 1250 °C. The same coating system was tested in one engine used by Texaco in Bakersfield, CA (USA), which successfully endured 14,000 h of service, leading to a significant reduction in the emission levels of NOx and CO. Nevertheless, unacceptably high recession rates were measured for the BSAS top-coat at temperatures above 1400 °C in combustion environments ($p = 6$ atm, $p_{H2O} = 0.6$ atm, $v = 24$ m s^{-1}). Another key issue of this coating system was the reaction of the BSAS top-coat with SiO_2 arising from the oxidation of the Si bond-coat, generating a low-melting glass ($T_{melt} \approx 1300$ °C) that caused the premature degradation of the EBC system [87].

2.5.1.3 3rd generation

In order to increase the stability of the developed EBCs to temperatures above 1400 °C, new top-coat materials that were compatible with the Si bond-coat and mullite or mullite + BSAS intermediate coating to replace the BSAS top-coat were investigated [19,22,83]. As shown in Section 2.4, especially rare-earth mono- and disilicates stand out due to the reduced CTE and low corrosion rates in hot gas environments up to 1450 °C. Despite the higher corrosion rates of the disilicates in comparison with monosilicates, the lower CTE results in a better thermomechanical compatibility to Si_3N_4, SiC and SiC/SiC substrates and should be considered when designing a reliable EBC for gas turbines. Since then, intensive research has been carried out during the development of the 3rd generation of coatings, which is based on rare-earth mono (RE_2SiO_5) and disilicates ($RE_2Si_2O_7$). Especially coatings based on the rare-earths lutetium (Lu) and ytterbium (Yb) are more commonly reported in the literature due to the reduced number of polymorphs and high stability in hot gas environments [19,25,30].

2.5.2 Rare-earth silicate environmental barrier coatings

Based on the high corrosion stability of the rare-earth mono- and disilicates in combustion environments, Lee *et al.* [22] investigated the thermomechanical stability of RE_2SiO_5 (RE = Sc, Yb and Lu) as a replacement for the BSAS top-coat from the 2nd generation for SiC, Si_3N_4 and SiC/SiC substrates. The coated specimens were prepared by air plasma spray (APS), followed by annealing at 1300 °C for 20 h in air to crystallize the silicate phases. Phase analysis of the coated samples indicated the volatilization of silica during the APS process, which resulted in the deposition of a residual phase composed of rare-earth oxides (RE_2O_3). The thickness of the coating systems corresponded to approximately 400 μm (100 μm top-coat and 300 μm bond-coat + intermediary coating). The coated samples were then thermal cycled with a high-power CO_2 laser in a flowing water vapor atmosphere (p = 1 atm, p_{H2O} = 0.9 atm v = 2.2 cm s^{-1}). The EBC surface temperature was set to 1380 °C to avoid the melting of the Si in the bond-coat at 1414 °C, which is considered to be an upper limit for its application. Each thermal cycle consisted of annealing the samples for 1 h at the desired temperature, followed by cooling to room temperature within 5 min. The coating systems resisted over 400 cycles without signs of spallation. However, in-thickness cracks were formed, which extended up to the bond-coat/substrate interface, caused by the large CTE mismatch between the different coating layers and the different phases present within the top-coat.

Richards *et al.* [10] deposited a tri-layer environmental barrier coating onto monolithic SiC substrates by air plasma spray. The coating microstructure consisted of a Si bond-coat (100 μm), a mullite intermediate coating (75 μm) and an Yb_2SiO_5 top-coat (75 μm). The authors report that the use of a mullite intermediary layer reduced the disparity between the CTE of the Yb_2SiO_5 top-coat and the monolithic SiC substrate, increasing the thermomechanical stability of the coated systems. After deposition, the coated substrates were annealed at 1300 °C for 20 h in air to crystallize the deposited layers. Yb_2SiO_5 and Yb_2O_3 were detected as main phases within the top-coat, which was explained by the loss of SiO_2 as SiO due to the high temperatures achieved by the plasma plumes during the coating process (2000-3500 °C). In addition, the crystallization step resulted in the formation of mudcracks, which extended up to the bond-coat due to the CTE mismatch between the different coating layers and between the Yb_2SiO_5 and Yb_2O_3 phases in the top-coat.

The same coating system was applied onto SiC/SiC substrates by air plasma spray, consisting of a Si bond-coat (133 μm), a mullite intermediate coating (183 μm) and an Yb_2SiO_5 top-coat (82 μm). In this work, Yang *et al.*[91] overcame the problem related to the volatilization of SiO_2 during the coating deposition and a pure phase and crack-free Yb_2SiO_5 top-coat was achieved. However, only limited information about the processing of the coating was provided. The stability of the coated substrates was tested during thermal cycling between 1200 and 200 °C in water vapor atmosphere (p = 1 atm, p_{O2} = 0.5 atm and p_{H2O} = 0.5 atm, flowrate 10 ml min^{-1}). The tests were carried out by annealing the coated samples at 1200 °C for 20 min and cooling to 200 °C within 20 min. After 700 cycles, the formation of horizontal and vertical cracks were seen. This had little influence on the coating adhesion, which reduced from 12.3 MPa to 6.8 MPa and only few spallation spots were seen.

A more stable coating system was achieved by Richards *et al.* [27], which consisted of a Si bond-coat (50 μm) and $Yb_2Si_2O_7$ top-coat (125 μm), deposited by air plasma spray onto SiC substrates annealed at 1200 °C in an Ar/H_2 atmosphere during deposition. The coated systems were then annealed at 1300 °C for 20 h in air, resulting in a dense, crack-free and crystalline microstructure. Due to the high temperatures achieved by the plasma plume and the consequent volatilization of SiO_2, a mixture of $Yb_2Si_2O_7$ and Yb_2SiO_5 was detected within the top-coat. The processed samples were submitted to thermal cycling tests in water vapor flowing atmosphere (p = 1 atm, p_{H2O} = 0.9 atm, v = 4.4 cm s^{-1}) by annealing at 1316 °C for 1 h and rapidly cooling to 110 °C within 10 min. The excellent thermomechanical compatibility between the coating layers and substrate was confirmed after 2000 cycles, thus only few spallation spots were seen at the edges of the coated system. During testing, the oxidation of the SiC substrate was avoided by the bond-coat, forming a SiO_2 thermally grown oxide layer (TGO). After thermal cycling tests a 15 μm thick and porous Yb_2SiO_5 layer was formed at the surface of the top-coat, arising from the corrosion of the $Yb_2Si_2O_7$ phase. Nevertheless, the coating system remained stable and only a few cracks were detected within its microstructure.

Bakan *et al.* [38] also investigated the deposition and hot gas corrosion behavior of a bilayer coating system consisting of a Si bond-coat and a $Yb_2Si_2O_7$ top-coat, which was applied by air plasma spray onto monolithic SiC substrates. To favor crystallization, the substrates were annealed in the temperature range of 1000-1100 °C during the coating deposition. An amorphous content of approximately 10 wt% was still detected within the coating layers even after carrying out an additional crystallization step at 1200 °C for 25 h in air. During the coating processing, the authors reported the volatilization of a great amount of SiO_2 from the $Yb_2Si_2O_7$ feedstock powder, which resulted in the deposition of a mixture of Yb_2SiO_5 and $Yb_2Si_2O_7$ phases with a maximum $Yb_2Si_2O_7$ content of 64 wt%. Despite achieving a dense microstructure with a total thickness of 250 μm (100 μm Si bond-coat and 150 μm $Yb_2SiO_5/Yb_2Si_2O_7$ top-coat), the large CTE difference between the two ytterbium silicate phases resulted in the formation of cracks in the top-coat after processing, which extended up to the Si bond-coat. The coated systems were subsequently tested in high velocity hot gas corrosion environments at 1200 °C for 200 h (p = 1 atm, p_{H2O} = 0.15 atm, v = 100 m s^{-1}). In comparison to uncoated SiC substrates, the coating system effectively hindered mass loss and a slight mass gain was measured after 200 h, indicating the predominance of oxidation reactions. The corrosion of the top-coat led to the formation of a porous Yb_2SiO_5 scale. The authors concluded that the mass gain was caused by the oxidation of the Si bond-coat, which led to the formation of a SiO_2-TGO.

Ueno *et al.* [41] compared different techniques for the deposition of $Lu_2Si_2O_7$ coatings on Si_3N_4 substrates, which were prepared by physical vapor deposition (PVD) and the reaction sintering method. The PVD method was efficient for the processing of dense coatings, however the deposition rate was estimated to be 1 μm h^{-1}. The reaction sintering was carried out by pre-oxidizing the Si_3N_4 substrates at 1100 °C for 48 h to form a dense SiO_2-layer and embedding them in a Lu_2O_3 power bed followed by sintering at 1500 °C for 2 h in Ar atmosphere (6 atm), allowing the developed SiO_2 to react with Lu_2O_3 to obtain $Lu_2Si_2O_7$. Among the tested techniques, the authors concluded the reaction sintering method to be the most efficient due to possible coating of complex geometries, the short coating time and low costs.

In further work, Ueno *et al.* [92] used the reaction sintering method for the deposition of a $Lu_2Si_2O_7$ bond-coat on turbine grade Si_3N_4 - SN282. Thereafter, a $Lu_2Si_2O_7$/mullite intermediary layer and $Lu_2Si_2O_7$ top-coat were deposited by air plasma spray, achieving a total thickness of 380 µm. The composition of the top-coat consisted of a mixture of Lu_2SiO_5 and $Lu_2Si_2O_7$ due to the volatilization of SiO_2 during the coating processing. The recession behavior of the coatings was evaluated by hot gas corrosion at 1300 °C for 500 h. In this case, steam generated at 200 °C was directly sprayed onto the surface of coated samples at a flow speed rate of 50 m s^{-1}. The corrosion of the surface of the $Lu_2Si_2O_7$ top-coat led to the formation of a porous Lu_2SiO_5 phase. Despite the thick and dense coating microstructure, the oxidation of the Si_3N_4 substrate could not be hindered. This resulted in the formation of a SiO_2-TGO and consequently cracks at the interface between Si_3N_4/coating system.

From the same research group, Jayaseelan *et al.* [46] deposited $Lu_2Si_2O_7$ coatings on Si_3N_4 by the sol-gel technique. The authors prepared a sol-gel precursor based on $Lu(NO_3)_3 \cdot nH_2O$ and tetraethyl orthosilicate (TEOS) in 2*n*-propanol, followed by the coating of Si_3N_4 substrates. A sintering step at 1550 °C for 4 h in air yielded a 20 µm thick, dense and crystalline $Lu_2Si_2O_7$ layer. The authors report that the advantage of this technique was the easy deposition of coatings onto complex geometries. However, no detailed information on the deposition method and the microstructure of the coating was provided.

A similar coating strategy was reported by Chen and Klemm [55] for the deposition of ytterbium mono- and disilicates onto Si_3N_4 based on a polysiloxane and Yb_2O_3. The coating processing occurred by dipping the Si_3N_4 substrate into a polysiloxane/Yb_2O_3 slurry, followed by pyrolysis at 1000 °C in air and sintering at 1500 °C in N_2 atmosphere, achieving a thickness of 20 μm. However, no information on the crystalline phase composition was provided. The coated Si_3N_4 substrates were tested against hot gas corrosion at 1450 °C for 200 h (p = 1 atm, p_{H2O} = 0.18 atm and v = 100 m s^{-1}), which resulted in the rapid spallation of the coatings due to the high residual porosity content and insufficient adhesion to the substrate.

2.5.3 Summary

First tests with mullite EBCs for non-oxide silicon-based substrates as Si_3N_4, SiC and SiC/SiC revealed their insufficient stability in combustion environments, leading to high recession rates. The volatilization of $Si(OH)_4$ from mullite led to the formation of Al_2O_3, which resulted in a large CTE mismatch to the used substrates, and in the rapid spallation of the coatings. The same problem was reported for the coatings of the 1st generation, which consisted of an yttria-stabilized zirconia top-coat applied onto the mullite coating.

In the 2nd generation, the deposition of a corrosion resistant coating system and a suitable oxidation protection of the substrate was achieved by replacing the coatings of the 1st generation by a Si bond-coat and a BSAS top-coat (1-xBaO-xSrO-Al_2O_3-2SiO_2, $0 \leq x \leq 1$). Field tests were carried out with coated ceramic components by Solar Turbines and Texaco over 24,000 h without failure at a maximum firing temperature of 1250 °C, thus reducing the emission levels of NOx and CO. Nevertheless, this coating system showed a limited stability at temperatures above 1300 °C due to the rapid recession rate and the formation of a eutectic phase by the reaction of BSAS with SiO_2 arising from the oxidation of Si in the bond-coat.

In the 3rd generation, research focused on materials to endure service at temperatures beyond 1400 °C, which led to the development of rare-earth silicate-based systems. Due to the reduced number of polymorphs and high stability in hot gas environments, coatings with lutetium (Lu) and ytterbium (Yb) silicates are more commonly reported in the literature [19,22,25,30]. In special, two coating systems have proven their potential for application in combustion environments. The first one consists of a Si bond-coat, a mullite intermediary layer and a rare-earth monosilicate (RE_2SiO_5, RE = Yb or Lu) top-coat. The second system is based on a Si bond-coat and a rare-earth disilicate ($RE_2Si_2O_7$) top-coat. Fig. 2.5.1 depicts a representative schema of the promising coating systems of the 3rd generation.

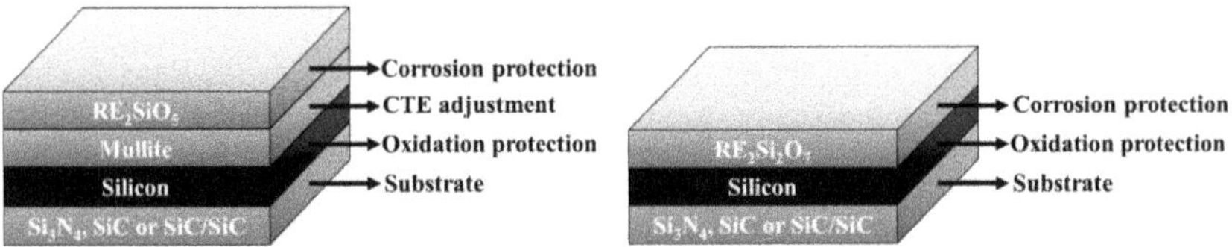

Fig. 2.5.1 Representative schema of the environmental barrier coating systems of the 3rd generation for protection of non-oxide silicon-based ceramics in combustion environments.

In both systems, the Si bond-coat was reported to increase the adhesion strength due to the chemical compatibility with non-oxide silicon-based ceramics, besides functioning as a barrier against the diffusion of oxygen, thus protecting the substrate during long-term service. In comparison with the monosilicate, the disilicate phase exhibits a higher corrosion rate [19]. However, the lower CTE of the disilicates (e.g. $Yb_2Si_2O_7$, CTE = 3.91 10^{-6} K^{-1} [19]) results in a better thermomechanical compatibility to the Si bond-coat (CTE = 4.1 10^{-6} K^{-1} [27]) and the Si_3N_4, SiC and SiC/SiC substrates (e.g. Si_3N_4, CTE = 3.60 10^{-6} K^{-1}). Therefore, a mullite intermediary layer (CTE = 5.71 10^{-6} K^{-1} [19]) is necessary to avoid the large CTE mismatch between the monosilicate (e.g. Yb_2SiO_5, CTE = 6.65 10^{-6} K^{-1} [19]), the Si bond-coat and the substrate in the first coating system. Nevertheless, the higher complexity of the tri-layer EBC often resulted in the formation of a largely cracked microstructure already after the coating deposition or during thermal cycling [10,27].

Typical deposition methods for Lu and Yb-derived EBC systems consist primarily of air plasma spray (APS) [10,22,27,38,91,92], followed by PVD [41], reaction sintering [41,92], the sol-gel route [46] and polymer derived ceramics [55]. In contrast to the other techniques, thick and dense coating systems with over 400 μm in thickness can processed by APS, which is desired for corrosion protection, considering the slow and continuous degradation of the coatings in combustion environments. Nevertheless, the greatest disadvantages of this method consist in the costly procedure, the difficult coating of complex geometries, the volatilization of SiO_2 from the used feedstock powders when depositing the coating and the additional crystallization step required post deposition. Due to the high temperatures achieved by the plasma plume during the coating process, SiO_2 volatilizes from the mullite and rare-earth silicate feedstock powders, which leads to the deposition of mixed silicate or residual oxide phases. This results in *in situ* CTE mismatches and the consequent formation of cracks. Moreover, the rapid cooling rates of the molten feedstock powders during the deposition of the coating results in an amorphous microstructure, which requires a long additional crystallization step at temperatures of up to 1300 °C for over 20 h in air. The main disadvantages of the deposition of coatings by PVD are the coating of complex geometries and the low deposition rates, requiring long processing times. In contrast, the deposition of coatings by the reaction sintering method, sol-gel route or polymer derived ceramics offer a promising and cost reduced approach for the deposition of coatings onto complex geometries. Another major advantage of these methods is the deposition of fully crystalline coatings with a controlled composition.

2.6 Polymer derived ceramics

In technical applications such as in gas turbines, surface protection with corrosion resistant coatings based on rare-earth silicates can prolong the lifetime of ceramic components at high temperatures in these harsh environments [19,22]. These ceramic coatings are typically applied by various techniques such as air plasma spray (APS) [10,22,27,38,91,92], physical (PVD) [20,41] or chemical vapor deposition (CVD) [39,40]. High costs, scale up problems and the difficult coating of complex geometries are crucial disadvantages of these techniques. An alternative approach is to employ preceramic polymers. Their flexibility in form enables the deposition of ceramic coatings by simple polymer processing techniques such as dip- or spray-coating, which are suitable for both large volumes and complex geometries. In this regard, the polymer derived ceramics (PDC) route is mainly based on reactive silicon-containing preceramic polymers from the binary systems SiN (polysilazanes), SiC (polycarbosilanes) and SiO (polysiloxanes), SiB (polyborosilanes) and their combination as the ternary SiCN, SiCO systems as well as quaternary SiCNO, SiBCN and SiBCO systems.

Commonly, hydrogen, aliphatic or aromatic organic side groups are attached to the silicon atom, which influence their cross-linking behavior, physicochemical properties and control the pyrolysis chemistry. A heat treatment usually at temperatures between 600-800 °C converts the preceramic polymers into stable ceramic materials with attractive properties for application in harsh environments at high temperatures as for example high corrosion and oxidation stability, high hardness and creep resistance. Depending on the treatment atmosphere, different ceramic yields and compositions are achieved [42,93].

Among the precursors suitable for coating application, silazanes, with Si and N in the chemical backbone, stand out due to the commercial availability, high ceramic yield, strong adhesion to most substrates and high stability in oxidative and corrosive media, besides the chemical compatibility to non-oxide silicon-based ceramics [42,43,47]. To better understand the cross-linking and pyrolysis behavior of the developed coating system during pyrolysis in air, which contains the perhydropolysilazane Durazane 2250 and the silazane Durazane 1800, this section emphasizes the polymer to ceramic transition of silazanes and examples of silazane-based ceramic coatings.

2.6.1 Polymer to ceramic transition

The polymer to ceramic transition of preceramic polymers as silazanes occurs upon thermolysis, which is usually divided in cross-linking, ceramization and crystallization. Cross-linking occurs at low temperatures from 100 to 400 °C, which is responsible for the transformation of the preceramic polymers into infusible materials or thermosets. Thereby, the volatilization of low molecular weight oligomers is hindered, which leads to higher ceramic yields and allows the processed materials to retain the shape during the ceramization process as long as it occurs without melting [93]. The cross-linking of polysilazanes can be either achieved thermally or induced by the use of initiators or catalysts. With the appropriate side groups, cross-linking can occur by hydrosilylation, vinyl polymerization, dehydrocoupling and transamination [93–95].

Hydrosilylation reaction occurs in oligosilazanes containing Si-H and vinyl groups in the temperature range between 100 and 120 °C, forming Si-C-Si and Si-C-C-Si bonds. Dehydrocoupling involves Si-H and N-H groups or two Si-H groups, where Si-N and Si-Si bonds form, respectively, along with the evolution of H_2 at temperatures of about 300 °C. Vinyl polymerization is reported to occur at approximately 170 °C and involves no mass loss. Nevertheless, the addition of suitable radical initiators as dicumyl peroxide reduces the temperature and time necessary for cross-linking, which can be carried out at temperatures lower than 130 °C [96,97]. At last, transamination takes place in the temperature range from 200 to 400 °C and is associated with the evolution of volatile amines, ammonia and oligomeric silazanes, thus leading to mass loss and a decreased nitrogen content in the final ceramic material. Alternatively, cross-linking of silazanes can also be induced in air. The high sensitivity of the reactive Si-H groups and Si-N bonds of silazanes in the presence of oxygen or moisture from the atmosphere leads to its incorporation and the elimination of NH_3 and hydrogen as H_2 and H_2O, following the reaction mechanisms from Eq. 2.6.1 to 2.6.3 [47,93].

$$2 \equiv \mathrm{Si} - \mathrm{H} + \mathrm{O_2} \rightarrow \equiv \mathrm{Si} - \mathrm{O} - \mathrm{Si} \equiv + \mathrm{H_2O} \qquad \text{(Eq. 2.6.1)}$$

$$\equiv \mathrm{Si} - \mathrm{NH} - \mathrm{Si} \equiv + \mathrm{H_2O} \rightarrow \equiv \mathrm{Si} - \mathrm{O} - \mathrm{Si} \equiv + \mathrm{NH_3} \qquad \text{(Eq. 2.6.2)}$$

$$2 \equiv \mathrm{Si} - \mathrm{H} + 2\,\mathrm{H_2O} \rightarrow \equiv \mathrm{Si} - \mathrm{O} - \mathrm{Si} \equiv + 2\,\mathrm{H_2} + \mathrm{H_2O} \qquad \text{(Eq. 2.6.3)}$$

Motz *et al.* [47] report that the incorporation of oxygen occurs up to approximately 400 °C, followed by the formation of a dense and passivating SiO_2 layer, which inhibits the further oxidation of the SiCN system. By annealing at higher temperatures, the ceramization of the cross-linked preceramic polymer starts, characterized by the volatilization of organic groups as CH_4 or the related oxidation products, depending on the treatment atmosphere, and leads to the formation of amorphous ceramics of the systems SiN(O) or SiCN(O) up to 1000 °C [42,47,93].

If the carbon fraction present in the starting precursor is high enough, part of it segregates during pyrolysis, creating a free carbon phase. Further heating induces the rearrangement of the amorphous network, and crystallization starts at temperatures above 1200 °C, becoming more pronounced at above 1400 °C. In protective atmospheres, this step leads to the further elimination of N_2 and the formation of Si_3N_4 and SiC crystalline domains [93,98]. In contrast, the formation of a dense and bonded oxide scale during processing in air delays the crystallization of SiCNO systems. This is explained by the further increase in the nitrogen partial pressure in the interior of the pyrolysates, hindering the elimination of N_2. Thus, polysilazanes pyrolyzed in air exhibit a thermal stability of up to 1600 °C, forming then crystalline SiO_2, SiC and Si_3N_4 phases [47].

The polymer to ceramic transition of polysilazanes is accompanied by a large volume shrinkage and a pronounced density increase. The density typically increases by a factor of 2 to 3 from the precursor (1.1 g cm^{-3}) to the crystalline ceramic residue (SiO_2, 2.2 to 2.6 g cm^{-3}, Si_3N_4 and SiC, 3.0 to 3.2 g cm^{-3}), which is accompanied by a volume shrinkage that exceeds 50%. When these volume changes cannot be relaxed by viscous flow or diffusion process, extensive cracking and pore formation generally compromise the integrity of the processed components. Hence, the conversion of a polymer compact to a dense ceramic compact is difficult to achieve [48].

The shrinkage and densification also limits the thickness of dense, crack-free and well-adherent pure silazane coatings to less than 3.5 μm [49,50,99], whereas thicker coatings are required for application in very harsh corrosive environments. Intrinsic shrinkage and porosity formation during pyrolysis can be hindered by the addition of fillers, thus reducing the fraction of the shrinking polymer phase. Thereby, the polysilazane precursor may act as a binder and as a reaction partner to the filler particles. These fillers are usually classified as passive, active and meltable.

Passive fillers (e.g. Al_2O_3, Si_3N_4 and ZrO_2) are inert to the system, undergoing no mass or volume change during pyrolysis, disregarding thermal expansion. In contrast to passive fillers, active fillers (e.g. Si, silicides and metal powders) undergo reaction with the atmosphere, precursor or pyrolysis products, thus expanding in volume and yielding tailored ceramic phases. Meltable fillers (e.g. glasses, Si or Si-alloys) may or may not react with the other components of the coating system, but their most important function is to melt during pyrolysis, filling cracks and reducing thermal stresses within the coatings due to CTE mismatches at high temperatures [42,48,93]. Consequently, the deposition of dense coatings of up to 100 μm for protection of steel substrates against oxidation and corrosion up to 700 °C was demonstrated [51].

2.6.2 Silazane-based ceramic coatings

As mentioned in Section 2.6.1, the large shrinkage and densification of polysilazanes limit the application of crack-free and dense unfilled coatings to only a few micrometers in thickness. This drawback can be overcompensated by the addition of suitable fillers, hence reducing the fraction of the shrinking polymer phase and enabling the deposition of thicker coatings with tailored properties [48,51]. Regarding this strategy, different coating systems were developed for ceramic and steel substrates based on polysilazanes.

Bill and Heimann [100] developed a protective coating based on a polysilazane and Si to enhance the oxidation stability of carbon fiber reinforced SiC composite (C/C-SiC). Therefore, elemental Si particles were dispersed in a toluene solution containing the commercially available polysilazane NCP 200. The C/C-SiC substrates were then dip-coated in the coating slurry inside a glovebox and pyrolyzed at 1100 °C in protective atmosphere. A coating thickness of approximately 20 μm was achieved and an adhesion strength of up to 14 MPa was measured. The coated substrates were tested against oxidation up to 1260 °C in air. Despite the enhanced durability of the coated C/C-SiC substrates, the coating system exhibited a porous and cracked microstructure that enabled the elimination of the carbon fibers after annealing for prolonged times at 1260 °C in air.

Bakumov *et al.* [101] synthesized silver nanoparticles with an average particle size of 5-7 nm, which were dispersed in a VL20 polysilazane solution. Thin films were applied by dip- and spin-coating onto steel and quartz substrates, followed by cross-linking at 290 °C and pyrolysis in the temperature range of 700 - 1000 °C in nitrogen atmosphere. After pyrolysis, the coating exhibited a homogeneous SiCN microstructure with silver nanoparticles embedded in it. Thereby, the resulting properties of the coatings were tailored, demonstrating a very high bactericide activity. The authors concluded that the simplicity and versatility of polymer derived ceramic shaping techniques combined with the very high anti-bacterial activity of the silver nanoparticles is a suitable approach for the deposition of coatings aiming application in biomedical, food and other industries.

In 2011, Günthner *et al.* [51] developed a double layer polysilazane-based environmental barrier coating for steel. In this coating system, PHPS was applied as a bond-coat due to its outstanding oxidation protection and excellent adhesion to steel substrates. The top-coat consisted of an organosilazane, ZrO_2 as passive filler and meltable glass powders as densifying agents. Pyrolysis was carried out at 800 °C in air, resulting in a well-adherent and dense coating layer with thicknesses of up to 100 μm. Cyclic oxidation tests were then carried out with coated samples between 700 °C and room temperature with heating and cooling rates of 10 K min^{-1} and an annealing time of 30 min at 700 °C. After three cycles the coating system remained crack-free and well-adhered to the steel substrate, effectively protecting it against oxidation.

Wang *et al.* [52] compared the performance of three different coating systems for protection of Inconel 617 superalloy substrates against static oxidation at 800 °C for up to 200 h in air. The three coatings systems consisted of (i) PHPS, (ii) a polysiloxane and $ZrSi_2$ as active filler and (iii) a combination of both systems, whereby the silazane PHPS was applied as the bond-coat. The different coating systems were applied by dip-coating followed by pyrolysis at 800 °C for 2 h in air. The unfilled PHPS coating (system i) achieved a thickness of 1.5 μm, whereas the deposition of the developed polysiloxane/$ZrSi_2$ slurry resulted in coatings with an increased thickness of up to 25 μm (systems ii and iii). Static oxidation at 800 °C for 200 h in air showed the significant reduction of the oxidation rate by all three coating systems, in comparison with uncoated substrates. The authors reported the system (iii): PHPS bond-coat and polysiloxane and $ZrSi_2$ top-coat, to be the most suitable for the protection of Inconel 617 against oxidation, by acting as an effective diffusion barrier against oxygen.

Günthner *et al.* [53] developed coatings for protection of steel substrates against oxidation and corrosion up to 700 °C. The authors used a perhydropolysilazane (PHPS) from the binary system SiN and BN as a passive filler. The coatings were applied by dip-coating, followed by pyrolysis at 800 °C in air. The resulting coating system exhibited a thickness of 12 μm, was dense and well-adhered to the steel substrate. The excellent adhesion strength was attributed to the formation of Fe-O-Si bonds between NH groups of the silazane and OH groups adsorbed at the surface of the steel substrate, besides the diffusion of elements from the substrate into the PHPS-based coating during pyrolysis. In comparison with uncoated substrates, a significantly reduced mass gain was measured for the coated samples after oxidation at 700 °C for 10 h in air, which relied on the formation of a dense SiO_2 layer from PHPS, slowing the diffusion of oxygen through the developed coating system.

Schütz *et al.* [102] investigated the stability of different polysilazane-based coatings on steel substrates against high temperature salt melt corrosion. The coatings were applied by dip- and spray-coating, followed by pyrolysis at 700 °C in air. Salt melt corrosion tests containing sodium, potassium and zinc chlorides and sulfates were performed at 530 °C up to 336 h. The best resulting systems consisted of a PHPS bond-coat and a top-coat composed of an oligosilazane, glass particles and ZrO_2, which also endured abrasive wear during field tests in waste incinerators for over 2 weeks under the same temperature conditions as the salt melt corrosion tests.

Barroso *et al.* [103] developed a novel thermal barrier coating for AISI 441 steel substrates based on a PHPS bond-coat and a top-coat consisting of an oligosilazane in combination with yttria stabilized zirconia (YSZ) as passive filler and $ZrSi_2$ as active filler. The bond-coat was applied by dip-coating, followed by cross-linking at 500 °C in air, before the deposition of the top-coat by the doctor blade technique. The coating system was pyrolyzed at 1000 °C for 1 h in air, achieving a thickness of 50 µm. Besides effectively protecting the substrate against oxidation during pyrolysis, the PHPS bond-coat was also reported to increase the adhesion strength of the coating system, which achieved value of 10 MPa. Due to the high porosity content (27 vol%) and the thick coating microstructure, a thermal conductivity of 0.44 ± 0.04 W m^{-1} K^{-1} was measured, which is one of the lowest values ever published for YSZ-based coatings.

In 2016, Barroso *et al.* [104] also demonstrated the application of silazane-based coatings onto graphite dies used for the sintering of Si_3N_4 ceramics by hot-pressing. The coatings were based on an oligosilazane and hBN as passive filler, which were applied by spray-coating, followed by pyrolysis at 1000 °C in nitrogen atmosphere. The silazane-based coatings performed significantly better compared to the commonly used water-based hBN coatings. Hence, they were able to reduce the interactions of carbon species from the dies with the glassy additives in the Si_3N_4 ceramics, improving the quality of the final product and reducing the costs related to hard machining.

Tangermann-Gerk *et al.* [105] and Horcher *et al.* [106] developed a bilayer polysilazane-based coating system for steel substrates resistant to corrosion and abrasive wear. Therefore, a PHPS bond-coat was applied by dip-coating, followed by the spraying of the top-coat slurry, containing an oligosilazane, glass powders and ZrO_2. Pyrolysis was carried out by employing a Nd:YAG laser source, resulting in a fully dense and compact microstructure with a thickness of up to 20 μm. The coating system exhibited an increased adhesion strength to the steel substrate of 23.8 MPa and a very high hardness of 12.75 GPa. Wear tests with the coated samples proved the good abrasion resistance of the laser pyrolyzed coatings. Moreover, no signs of degradation were seen after continuously spraying the coated samples with a 5% saline solution aerosol for 168 h in a chamber kept at 35 °C.

In summary, the excellent adhesion and the potential of silazane-based coatings for protection of ceramic and metal substrates against oxidation and corrosion in harsh environments has been proven. Nevertheless, their protective effect often relies on the formation of a passivating SiO_2-scale, which is rapidly degraded in combustion environments. Thus, in order to enable the application of silazane-based coatings in gas turbines, the conversion of forming SiO_2 into corrosion resistant phases is imperative.

2.7 Synthesis of rare-earth silicates

The conversion of SiO_2 into corrosion resistant rare-earth silicates can be achieved by the solid-state reaction with rare-earth oxides (RE_2O_3), thus yielding rare-earth mono- (RE_2SiO_5) or disilicates ($RE_2Si_2O_7$), according to molar ratio between SiO_2 and RE_2O_3 (Eq. 2.7.1 and 2.7.2).

$$RE_2O_3 + SiO_2 \rightarrow RE_2SiO_5 \qquad \text{(Eq. 2.7.1)}$$

$$RE_2SiO_5 + SiO_2 \rightarrow RE_2Si_2O_7 \qquad \text{(Eq. 2.7.2)}$$

In this regard, Aparicio and Durán [107] prepared yttrium monosilicate from SiO_2 and Y_2O_3 powders and studied its densification behavior. The SiO_2 and Y_2O_3 powders were mixed in a mol ratio of 1, pressed into bulk monoliths and sintered up to 1600 °C for 100 h. At above

1100 °C, the monosilicate phase started forming, resulting in the shrinkage of the pressed monolith. A maximum density corresponding to 71% of the theoretical density of the Y_2SiO_5 phase was reached after sintering at 1600 °C for 30 h in nitrogen atmosphere. The authors still report that sintering was strongly increased by the addition of 3 mol% Al_2O_3, reaching a density of over 90% the theoretical one at 1600 °C within the same time.

Nasiri *et al.* [108] synthesized Y, Gd, Er, Yb and Lu monosilicates through the reaction of the corresponding oxides and SiO_2 in a mol ratio of 1. The starting materials were uniaxially pressed at 50 MPa and sintered at 1580 °C for up to 12 h in air. After sintering, relative densities of above 94% were obtained for the pressed monoliths in regard with the respective monosilicate phases. Nevertheless, the resulting ceramics contained a large amount of unreacted SiO_2 and oxide phases of up to 15 wt%.

Maier *et al.* [109,110] investigated the synthesis of Y, Lu, Yb and Gd disilicates from SiO_2 and the corresponding oxides in a molar ratio of 2 (SiO_2:RE_2O_3) in the temperature range between 1300 and 1600 °C. Rare-earth monosilicates were predominantly detected at already 1320 °C. By increasing the annealing time to 48 h at this temperature or heating at above 1400 °C, phase pure rare-earth disilicates were formed. Nevertheless, no information on the resulting microstructure or density was provided.

Wang *et al.* [111] adopted a similar strategy for the synthesis of single-phase ytterbium disilicate (β-$Yb_2Si_2O_7$). The authors used a mixture of Yb_2O_3 and a SiO_2 gel derived from tetraethoxysilane (TEOS) in a theoretical SiO_2:Yb_2O_3 mol ratio of 2 (system 1). For a matter of comparison, the authors carried out the same reaction between Yb_2O_3 and SiO_2 commercial powders by the traditional ceramic powder route (system 2). The two systems were pressed at 30 MPa and fired at temperatures ranging from 1000-1550 °C in air. By using TEOS as a source of SiO_2 the authors were able to reduce the temperature for the synthesis of pure phase β-$Yb_2Si_2O_7$. Hence, after pyrolysis at 1550 °C for 4 h in air, only β-$Yb_2Si_2O_7$ was detected, whereas considerable amounts of unreacted Yb_2SiO_5 were still detected for system 2 under the same conditions. The β-$Yb_2Si_2O_7$ phase resulting from pyrolysis of system 1 at 1550 °C was subsequently milled and uniaxially pressed at 240 MPa, followed by sintering at temperatures ranging from 1300 to 1550 °C for 4 h in air. At 1300 °C, an open porosity of 33.4 vol% was measured, corresponding to a density of 3.96 g cm^{-3}. At above 1300 °C, the pressed monoliths started to densify, reaching an open porosity content of 11.3 vol% and a bulk density of 5.25 g cm^{-3} at 1500 °C. Nevertheless, only after sintering at 1550 °C for 4 h a dense microstructure could be achieved. Thereby, the open porosity content decreased to 0.6 vol%, resulting in a bulk density of 5.71 g cm^{-3}, which corresponded to approximately 93% of the density of the β-$Yb_2Si_2O_7$ phase (6.15 g cm^{-3}).

Alternatively, rare-earth silicates have also been synthesized by the polymer derived ceramic route from mixtures of polysiloxane resins and RE_2O_3. In this way, Bernardo *et al.* [112] prepared slurries containing Y_2O_3 nanopowders with a mean particle size ranging from 30-50 nm and a commercial polysilsesquioxane in the respective mol ratios to obtain yttrium mono- (Y_2SiO_5) and disilicates ($Y_2Si_2O_7$). The slurries were subsequently dried and milled into homogeneous powders, followed by pyrolysis in the temperature range from 900 to 1500 °C for 1 h in air. The silicate phases started forming at above 1000 °C, reaching a stoichiometric conversion into the respective Y_2SiO_5 and $Y_2Si_2O_7$ phases at 1400 °C.

Liu *et al.* [113] adopted a similar strategy for the synthesis of yttrium silicates by employing different weight ratios between the selected polysiloxane and the Y_2O_3 powder with a mean particle size of 1-2 μm. Thereafter, slurries were prepared with both components to obtain a homogeneous dispersion, which were dried at 70 °C for 10 h and subsequently milled. The resulting fine powders were then pyrolyzed at 900 °C for 2 h in air, followed by a heat treatment at 1500 °C for 5 h under flowing argon, forming yttrium mono- and disilicates. The synthesized silicate powders were applied as coatings to 2D C/SiC composites by chemical vapor infiltration, followed by sintering at 1400 °C for 5 h in argon atmosphere to obtain a dense microstructure. The coated specimens were tested in a water vapor corrosion environment ($p = 1$ atm, $p_{O2} = 0.5$, $p_{H2O} = 0.5$, $v = 1.04\ 10^{-3}$ m s^{-1}) at 1400 °C for up to 200 h. In comparison with uncoated substrates, no mass change was detected, demonstrating the efficiency of the developed yttrium silicate-based coatings in wet environments for protecting the C fibers against oxidation.

Chen and Klemm [55] prepared *in situ* generated ytterbium silicate coatings for Si_3N_4 by employing a mixture of Yb_2O_3 and a polysiloxane in the corresponding mol ratios to obtain ytterbium mono- and disilicates. The coating processing occurred by dipping the Si_3N_4 substrate into a polysiloxane/Yb_2O_3 suspension, followed by pyrolysis at 1000 °C in air and sintering at 1500 °C in N_2 atmosphere. However, no information on the resulting crystalline phase composition was provided. The coated Si_3N_4 substrates were tested against hot gas corrosion at 1450 °C for 200 h ($p = 1$ atm, $p_{H2O} = 0.18$ atm and $v = 100$ m s^{-1}). Due the residual porosity and insufficient adhesion to the substrate, the coatings rapidly spalled-off during testing.

The sensibility of silazanes precursors to moisture in the atmosphere is usually a disadvantage in combustion environments [15,16,114]. However, it was shown that the high reactivity of NH groups from the silazanes with OH groups adsorbed at the surface of metal and ceramic substrates promotes the adhesion of coating systems [50]. In addition, thermal treatment in air is favorable for the cross-linking of silazane-based coatings by the formation of Si-O-Si bonds, yielding SiO_2 during pyrolysis [47,49], which is necessary for the conversion of rare-earth oxides in silicate phases. Based on this premise, Seifert *et al.* [115] investigated the formation of magnesium silicates by the reaction of perhydropolysilazane (PHPS) with MgO during pyrolysis in air. After pyrolysis at 1100 °C in air, $MgSiO_3$ and Mg_2SiO_4 were already main crystalline phases, formed by the solid-state reaction of the polysilazane with MgO, whereas only traces of the unreacted oxide were detected.

By comparing the different polymer derived ceramic precursors employed for the synthesis of rare-earth silicates, it is expected that a better chemical compatibility to non-oxide Si-based ceramics and a stronger adhesion can be achieved by the use of silazanes instead of siloxanes for coating application [10,27,37]. Moreover silazanes stand out due to the commercial availability, high ceramic yield, and outstanding stability in oxidative and corrosive media [42,43,47].

2.8 Preliminary studies: silazane and Yb_2O_3-based systems

It is known that the protective effect of silazane-based coatings against corrosion and oxidation in harsh environments often relies on the formation of a passivating SiO_2-scale, which undergoes detrimental reaction with water vapor in combustion environments and is rapidly corroded at temperatures above 1200 °C. The basic development of the $Yb_2Si_2O_7$ coating consisted in understanding the role of silazanes in the conversion of Yb_2O_3 into ytterbium silicates upon pyrolysis. In preliminary studies, Yb_2O_3 was selected as an active filler in order to convert the forming SiO_2 from the silazane precursor Durazane 1800 into $Yb_2Si_2O_7$, increasing the stability of silazane-based coatings in combustion environments [54].

Monolithic samples were processed with Durazane 1800 and Yb_2O_3, whereas the molar ratio between Si from the precursor and Yb_2O_3 was varied in order to yield ytterbium mono- or disilicates within the temperature range of 1000 - 1500 °C in air. The formation of the respective silicate phases occurred at temperatures above 1200 °C, resulting from the solid-state reaction between the oligosilazane and Yb_2O_3. In comparison with similar Y_2O_3-based composites, the respective silicate phases were already formed at 1000 °C. The lower reactivity of Yb_2O_3-based systems was attributed to the larger mean particle size of the used Yb_2O_3 filler (d50 = 3.97 μm) in comparison with Y_2O_3 (d50 = 0.8 µm). Pyrolysis at temperatures above 1400 °C led to nearly the quantitative yield of ytterbium mono or disilicates, according to the corresponding molar ratio between Durazane 1800 and Yb_2O_3. Despite the considerable amount of carbon and nitrogen in the chemical structure of the oligosilazane, no related carbide or nitride phases were detected after pyrolysis in air. Moreover, the absence of precursor-derived SiO_2 indicated the potential of the adopted strategy into yielding corrosion stable ytterbium silicates. Nevertheless, a porosity content of up to 40 vol% was measured by He-pycnometry for the monolithic samples after pyrolysis, which represents a shortcoming regarding the development of environmental barrier coatings for protection of ceramic components in hot gas environments.

To evaluate the corrosion resistance of the prepared silicate phases, the corresponding Yb_2O_3-based composites were hot-pressed after pyrolysis in air, leading to fully dense monoliths. Additional samples were prepared by pyrolysis of the developed composites in argon atmosphere to study the influence of oxygen on the silicate phase formation and the corrosion resistance of the resulting phases. Hot gas corrosion tests at 1400 °C for 80 h (p = 1 atm, $p_{H2O} = 0.3$ atm, v = 1 mm s^{-1}) confirmed the stability of the dense silicate phases processed in air and only slight signs of degradation were detected. The characterization of the Yb-composites pyrolyzed in argon atmosphere revealed $Yb_4Si_2N_2O_7$ as the main crystalline phase before corrosion. The oxidation of this phase into ytterbium mono- and disilicates during corrosion and the considerable volume expansion associated with it led to the collapse of the samples. Thus, pyrolysis in air was proven to be imperative to obtain stable systems.

Coating experiments were then carried out to investigate the applicability of the developed composites onto Si_3N_4. Suspensions containing 25, 15, 7 and 4 wt% Durazane 1800 (with respect only to the mass of precursor and filler) were applied by spray-coating onto Si_3N_4 and subsequently pyrolyzed at 1400 °C for 1 h in air. In contrast to bulk monolithic samples, only a suspension with 4 wt% oligosilazane yielded stable coatings. It is well known that Durazane 1800 undergoes a significant shrinkage during pyrolysis [48,52]. In a free-standing sample, as the pressed monoliths, there are no constrains hindering this shrinkage. However, the adhesion of the coating to the substrate constrains its shrinkage, which may only occur freely across the thickness. Thus, a high fraction of silazane in the coating led to the generation of excessive stresses, resulting in cracking and spallation of the coatings. Fig. 2.8.1 (a) exhibits the SEM analysis of the cross-section of the coating containing 4 wt% Durazane 1800 and 96 wt% Yb_2O_3 after pyrolysis at 1400 °C for 1 h in air.

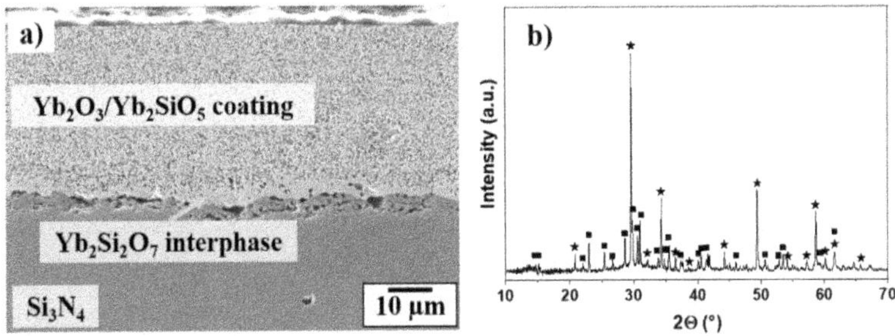

Fig. 2.8.1. (a) SEM micrograph of the cross-section of the Durazane 1800 and Yb_2O_3-based coating with 4 wt% silazane and respective (b) XRD analysis after pyrolysis at 1400 °C for 1 h in air (★ Yb_2O_3 (04-001-2438); ■ I2/a-Yb_2SiO_5 (00-040-0386)).

In Fig. 2.8.1 (a), the formation of a relatively thick and well-adherent layer with approximately 40 μm is evident. As evidenced by EDS analysis, the diffusion of silicon from the Si_3N_4 substrate into the coating enabled the formation of an ytterbium disilicate ($Yb_2Si_2O_7$) interface, which is desired due to the matching CTE with Si_3N_4 [19]. In contrast, the top layer is composed of the I2/a-Yb_2SiO_5 phase and unreacted Yb_2O_3, as detected by XRD (Fig. 2.8.1 (b)). The diffusion of Si from the substrate may still increase the adhesion of the coating system. Nevertheless, the coating microstructure is porous, which eases the permeation of moisture and oxygen, leading to the corrosion of the coated Si_3N_4 substrate. Furthermore, a low thermomechanical stability is expected due to the large CTE mismatch between the Yb_2O_3 (CTE = 8.4 10^{-6} K^{-1} [116]) and Yb_2SiO_5 (CTE = 6.65 10^{-6} K^{-1} [19]) phases and the Si_3N_4 substrate, which results in the *in situ* generation of stresses during thermal cycling, causing the rapid spallation of the coating. Therefore, the further densification of the coating and the conversion of the residual oxide and monosilicate phases were required for the development of stable environmental barrier coatings.

3 Experimental procedure

3.1 Materials

3.1.1 Substrates

3.1.1.1 Si_3N_4

Among the different oxide and non-oxide monolithic ceramics, Si_3N_4 is one of the most promising candidates for high temperature application. SNPU-Si_3N_4 (FCT Ingenieurkeramik GmbH, Germany) was chosen as the substrate for the development of the $Yb_2Si_2O_7$ coatings due to the commercial availability, extremely dense microstructure and low sintering additive content, consisting of a mixture of $Y_2O_3/MgO/Al_2O_3$. The mechanical properties of the selected Si_3N_4 are summarized in Table 3.1.1, according to the manufacturer [117].

Table 3.1.1. Properties of the selected Si_3N_4 substrate [117].

Properties	**Si_3N_4 substrate**
Density (g cm^{-3})	3.18 - 3.22
Residual porosity	< 0.5
Flexural strength at RT (MPa)	870
Weibull modulus	10 - 15
Hardness (GPa)	15.3
CTE (10^{-6} K^{-1})	3.2
Maximum service T (°C) (inert atmospheres)	1500
Maximum service T (°C) (oxidizing atmosphere)	1300

3.1.1.2 SiC/SiC

For the investigation of the transferability of the coating technology between non-oxide silicon-based ceramics, coating experiments were carried out with SiC/SiC substrates after the characterization of the coating system on Si_3N_4 substrates (Section 4.4). For this purpose, SiC/SiC composites with a fiber orientation of 0°/90° were obtained from BJS Composites GmbH, Germany. The open porosity content of the SiC/SiC substrates was estimated in 10.2 ± 1.0 vol% by the Archimedes' principle, and a density of 2.49 ± 0.03 g cm^{-3} was determined.

3.1.2 Coating slurries

Hereafter, a short description of the materials necessary for the development of the coating system is given, whereas more details about their selection will be provided in Section 3.2. The solvent di-n-butylether 99+% was acquired from Acros Organics BVBA, Belgium. Disperbyk-2070 was obtained from BYK-Chemie GmbH, Germany and used as a dispersing agent. The radical initiator Dicumyl peroxide (DCP) with 98% purity was purchased from Sigma-Aldrich Co. LLC., Germany. ZrO_2 milling beads were acquired from Sigmund Lindner GmbH, Germany. The silazane precursors Durazane 1800 and Durazane 2250 were obtained from Merck KGaA, Germany (Fig. 3.1.1). For the calculation of the volume fraction of the silazane precursors in the Si bond-coat and the Yb(2:1) top-coat slurries, a density value of 1.1 g cm^{-3} was used [48]. The fillers Yb_2O_3 (d50 = 3.97 μm) with 99.9% purity and elemental Si powder grade AX 05 (d50 = 3.04 μm) were purchased from Alfa Aesar GmbH & Co KG (Germany) and Kyocera K. K. (Germany), respectively. Table 3.1.2 summarizes some of the properties of the used Yb_2O_3 and Si fillers and the respective SEM micrographs are presented in Fig. 3.1.2.

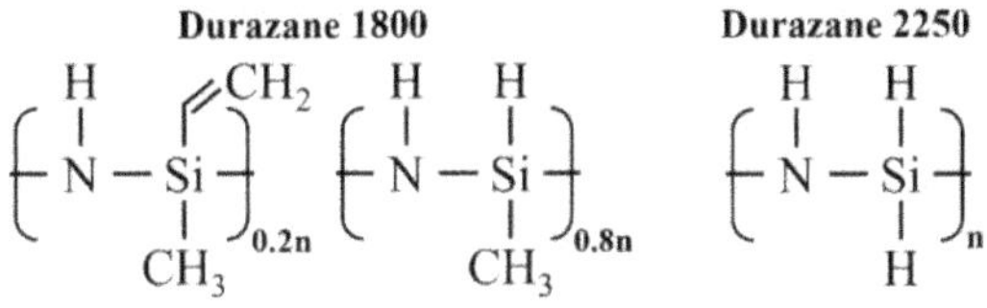

Durazane 1800 **Durazane 2250**

Fig. 3.1.1. Simplified structure of the oligosilazanes Durazane 1800 and Durazane 2250.

Table 3.1.2. General properties of the Yb_2O_3 and Si fillers.

Properties	**Yb_2O_3**	**Si**
Density (g cm^{-3})	9.17	2.33
Molar mass (Mw, g mol^{-1})	394.08	28.08
CTE (10^{-6} K^{-1})	8.4 [116]	4.1 [27]
Melting point (°C)	-	1414
d50 (μm)	3.97	3.04

Fig. 3.1.2. SEM micrographs of the selected (a) Yb_2O_3 and (b) Si fillers.

3.2 Composition and preparation of the coating slurries

3.2.1 Si Bond-coat

First $Yb_2Si_2O_7$ coatings on Si_3N_4 revealed the limited oxidation stability of the selected substrates at above 1400 °C in air due to the diffusion of sintering additives and impurities into the forming SiO_2 passivating layer, which leads to the reduction of its viscosity, increasing the oxidation rate [72,118,119]. The high temperatures necessary to densify the $Yb_2Si_2O_7$ coatings, as will be discussed in detail in Section 4.2, led to the enhanced oxidation of the selected substrates and the formation of blisters, compromising the mechanical stability of the deposited coatings. Therefore, the development of a bond-coat was necessary to prevent the oxidation of the substrate during pyrolysis.

For this purpose, the perhydropolysilazane Durazane 2250 (Fig. 3.1.1 (a)), available in a 20 wt% solution in di-n-butylether (DBE), was chosen due to the formation of a dense passivating SiO_2-scale to avoid the diffusion of oxygen. Moreover, its high reactivity with moisture favors the formation of chemical Si-O-Substrate bonds with adsorbed OH groups at the surface of metal and ceramic substrates, enhancing the adhesion of coating systems [49,50,120]. Elemental Si powder (Fig. 3.1.2 (b)) was added to increase the coating thickness, providing an improved oxidation protection of the substrate by the formation of SiO_2 and the densification of the bond-coat during pyrolysis in air.

To improve the adhesion of the coating system after pyrolysis and hinder the formation of blisters, different volume fractions between the components Durazane 2250 and elemental Si were tested, ranging from 90 to 70 vol% Durazane 2250 (20 wt% solution in DBE). A suitable bond-coat adhesion and the oxidation protection of the substrate were achieved with the slurry composition shown in Table 3.2.1. The suspension was prepared by stirring elemental Si in the 20 wt% Durazane 2250 solution in DBE.

Table 3.2.1. Composition of the Si bond-coat.

Components	Wt%	Vol%
Durazane 2250*	72	88
Si	28	12

* 20 wt% solution in DBE.

3.2.2 Yb(2:1) Top-coat

Besides the cross-linking via the formation of Si-O-Si bonds in air, the silazane Durazane 1800 can also react by dehydrocoupling and polymerization of the vinyl double bonds, leading to higher ceramic yields during pyrolysis. Therefore, this silazane was chosen for the development of the $Yb_2Si_2O_7$ top-coat, which was used as a binder as well as a source of silicon to convert the Yb_2O_3 filler (Fig. 3.2.1 (a)) into ytterbium disilicate.

During the investigation of the role of silazanes in the conversion of ytterbium oxide into silicates (Section 2.8), the low reactivity of this filler with Durazane 1800 was associated with the large mean particle size (d50 = 3.97 µm). Therefore, its surface area was increased by milling the purchased powder in a MiniCer (Netzsch GmbH, Germany) with ZrO_2 milling beads with a diameter of 1 mm, obtaining a powder with a d50 = 0.63 µm (Fig. 3.2.1 (b)), measured by laser granulometry (Granulomètre 850, Cilas-Alcatel S.A., France).

Fig. 3.2.1. SEM micrographs of the Yb_2O_3 filler used in the Yb(2:1) top-coat (a) Yb_2O_3 as purchased, (b) Yb_2O_3 after milling with 1 mm ZrO_2 beads.

Different molar ratios between Durazane 1800 and Yb_2O_3 can either lead to the formation of ytterbium mono or disilicates. However, the amount of silazane had to be restricted in first coatings based on the system Durazane 1800/Yb_2O_3, which resulted in a nonstoichiometric amount of Si to yield ytterbium disilicate as discussed in Section 2.8. Therefore, an additional filler was necessary to convert the residual Yb_2O_3 and Yb_2SiO_5 phases into $Yb_2Si_2O_7$ and also to compensate the shrinkage of the silazane during pyrolysis in air. Among the possible sources of SiO_2, the advantage of using elemental Si is the oxidation at relatively low temperatures of approximately 750 °C, which is accompanied by molar volume expansion by a factor of 2.2 and can overcome the above-mentioned drawbacks [71]. Hence, elemental Si powder (Fig. 3.1.2 (b)) was employed as an active filler and source of silica during pyrolysis in air to yield dense $Yb_2Si_2O_7$ coatings.

During the coating application, the evaporation of the solvent may lead to the faster settling of the suspended powder particles and cause the instability of the slurries. Di-n-butylether was chosen as the slurry solvent due to the elevated boiling point (141 °C) and low vapor pressure. The slurry was prepared by dispersing the fillers in di-n-butylether with help of 4.5 wt% of Disperbyk-2070 as dispersing agent, regarding the used filler mass. Disperbyk-2070 is suitable for the dispersion of pigment concentrates and it was selected due to the good solubility in di-n-butylether and the chemical compatibility with the precursor Durazane 1800. In a next step, a solution of Durazane 1800 and 3 wt% DCP was added to the mixture, which was stirred until a stable suspension was obtained. DCP acts as an initiator for polymerization and hydrosilylation reactions, promoting the cross-linking of Durazane 1800 at low temperatures and increasing the ceramic yield during pyrolysis [96].

The good dispersion of the Si and Yb_2O_3 fillers assures a balanced contact with Durazane 1800, thus enabling the homogeneous formation of silicate phases upon pyrolysis in air. ZrO_2 milling beads with a mean diameter of 2.0 mm were added to the slurry in order to achieve a higher level of homogeneity and decrease the mean particle size of the Si filler, consequently increasing its surface area and reactivity. As explained in Section 2.8 stable coatings from the Durazane 1800 and Yb_2O_3 system could only be obtained by limiting the amount of silazane to 4 wt%, corresponding to a volume fraction of approximately 25%. Based on the ceramic yield of the oligosilazane Durazane 1800 after pyrolysis in air at above 1000 °C (81 wt%) and the resulting elemental composition ($SiC_{1.15}N_{0.68}O_{0.42}$) [47], a total mol ratio of 2 Si:Yb_2O_3 was calculated to yield $Yb_2Si_2O_7$.

Different volume fractions of Durazane 1800 were tested, ranging from 25 to 12%. In one hand, the lowest volume fractions of Durazane 1800 resulted in an insufficient cohesion between the filler particles and a weak adhesion to the substrate, which led to the low mechanical stability of the coatings at room temperature. On the other hand, the highest volume fractions resulted in the spallation of the coating during pyrolysis, attributed to the high volume shrinkage of the oligosilazane [48]. A mechanically stable and well-adhered coating was achieved by adjusting the volume fraction of the silazane precursor to approximately 18%, according to the composition shown in Table 3.2.2, named Yb(2:1).

Table 3.2.2. Composition of the Yb(2:1) top-coat.

Components	Wt%	Vol%
Durazane 1800 + 3 wt% DCP	3	18
Yb_2O_3	86	55
Si	11	27

3.3 Processing of powder and monolithic samples

3.3.1 Processing of the Yb(2:1) composite powder

In order to acquire information on the formation of the $Yb_2Si_2O_7$ phase and its sintering behavior, any influence of the substrate was initially excluded by investigating powder and monolithic samples processed from the Yb(2:1) slurry. In this case, the solvent was removed from the slurry under stirring and reduced pressure (10^{-2} mbar) at 80 °C for 5 h, followed by milling the solid mixture in a Pulverisette 9 (Fritsch GmbH, Germany) for 60 s, yielding a homogeneous powder with a decreased particle size.

3.3.2 Cold pressing of Yb(2:1) monoliths

The processing of monoliths with approximate dimensions of Ø 12x3 mm occurred by uniaxially pressing the Yb(2:1) composite powder at 400 bar, followed by pyrolysis in air at the temperatures of 750, 900, 1000, 1100, 1200, 1300 and 1415 °C for 1 h in an HT04/18 chamber furnace (Nabertherm GmbH, Germany). To investigate microstructural and composition changes at increased annealing times and temperatures, two additional monoliths were pyrolyzed at 1415 and 1500 °C for 5 h in air.

3.4 Deposition of the Si/Yb(2:1) coating system

Before the coating application, the 30x25x5 mm Si_3N_4 substrates were cleaned in an acetone ultrasound bath for 25 min in order to remove any surface impurities, which may hinder the adhesion of the coating. The cleaned Si_3N_4 substrates were then dip-coated into the Si bond-coat slurry using a RDC 10 dip-coater (Bungard Elektronik GmbH & CO. KG, Germany) with a hoisting speed of 0.5 m min^{-1}. The coated substrates were dried and cross-linked at 400 °C for 2 h in air in order to yield a stable thermoset bond-coat. The Yb(2:1) slurry was then applied as the top-coat by spraying, using a 781S spray gun (Nordson Deutschland GmbH, Germany). Pyrolysis was carried out at 1415 °C for 5 h in air in an HT04/18 chamber furnace (Nabertherm GmbH, Germany) (see Section 4.2). Hereafter, the coating system was named **Si/Yb(2:1)**, referring to the **Si** bond-coat and the **Yb(2:1)** top-coat, respectively. Fig. 3.4.1 exhibits a scheme of the components and structure the developed coating system before pyrolysis.

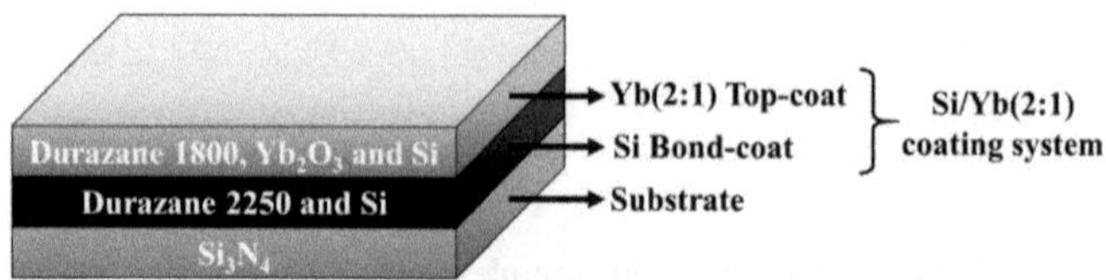

Fig. 3.4.1. Representative scheme of the components and structure of the Si/Yb(2:1) coating system before pyrolysis.

3.4.1 Deposition of thicker Yb(2:1) coatings

Since gas turbines are expected to operate for considerable amount of times and rare-earth silicate-based environmental barrier coatings are slowly corroded in hot gas environments, the deposition of thick coatings (> 100 μm) is desired to extend the life span of coated ceramic parts. The process optimization for the deposition of thicker Yb(2:1) coatings occurred after the characterization of the coating system resulting from Section 3.4. Thereby, a thickness of 68 μm was reached after pyrolysis at 1415 °C for 5 h in air (refer to Section 4.4).

After the application of the Si bond-coat according to Section 3.4, the deposition of thicker Yb(2:1) coatings occurred by performing successive spraying and intermediate pyrolysis cycles at 750 °C. In this regard, the deposition of a coating with 3 cycles consisted of 3 spraying cycles and 2 intermediate pyrolysis steps at 750 °C for 1 h in air. Following the last spraying cycle, the coating system was pyrolyzed at 1415 °C for 5 h in air.

3.5 Characterization methods

3.5.1 Pyrolysis behavior of the Yb(2:1) composite powder

In order to understand the pyrolysis behavior of the Yb(2:1) top-coat during the thermal treatment applied for the synthesis and sintering of the $Yb_2Si_2O_7$ phase, the composite powder resulting from the Yb(2:1) slurry was analyzed by thermogravimetry (TGA) in a STA 449 F5 Jupiter (Netzsch GmbH & Co. Holding KG, Germany), whereby the mass change as a function of the temperature was recorded. The fine-milled Yb(2:1) composite powder was heated up to a maximum temperature of 1415 °C at a rate of 5 K min^{-1} and a holding time of 5 h using synthetic air (80/20 vol% N_2/O_2, 50 ml min^{-1}). The resulting mass change was compared with the precursor Durazane 1800 + 3 wt% DCP and the fillers Yb_2O_3 and Si, measured under the same conditions.

3.5.2 Phase composition and microstructure of the Yb(2:1) monoliths

Details about the resulting microstructure and composition of the Yb(2:1) pressed monoliths after pyrolysis was analyzed by scanning electron microscopy (SEM) in a Gemini Sigma 300 VP (Carl Zeiss AG, Germany), coupled with an EDAX Octane Elect SDD, used for energy-dispersive X-ray spectroscopy (EDS). For cross-sectional analysis, the samples were sawn at the mid-section and embedded in an epoxy resin, which were subsequently grinded and polished up to 0.04 μm. Finally, the polished samples were coated with a 20 nm thin graphite layer, thus enabling a more precise distinction of the resulting phases after pyrolysis. The crystalline phase formation over the pyrolysis range between 750 and 1500 °C was investigated by X-ray diffractometry (XRD) in a D8 ADVANCE (Bruker AXS, Germany) using monochromatic CuKα radiation. For XRD analysis, the monoliths were grinded in a Pulverisette 9 mill (Fritsch GmbH, Germany) for 60 s to yield a more accurate signal from the crystalline phases within the respective samples. The resulting diffractograms were evaluated with the PDF-4+ 2020 structural database.

3.5.2.1 Volume change, porosity and density

In a reactive system as the Yb(2:1) monoliths, pyrolysis of Durazane 1800, the oxidation of the Si powder, the formation of ytterbium silicates and sintering effects affect the resulting microstructure and porosity content. In order to determine the volume change of the Yb(2:1) composite during pyrolysis in air, the apparent volume of the Yb(2:1) monoliths was compared before (V_o) and after (V') pyrolysis for each of the pyrolysis temperatures described in Section 3.3.2, according to Eq. 3.5.1. For this purpose, the respective dimensions were measured with a micrometer caliper and the volume was calculated.

$$\frac{\Delta V}{V_o}\,(\%) = \frac{(V' - V_o)}{V_o} \times 100 \qquad \text{(Eq. 3.5.1)}$$

The true volume of the Yb(2:1) monoliths after pyrolysis at different temperatures in air was measured by helium pycnometry in an Accupyc II 1340 (Micromeritics Instrument

Corporation, USA). The open porosity content (φ) was then calculated considering the difference between the apparent volume (V') and the true volume measured by He-pycnometry (V_{He}), according to Eq. 3.5.2.

$$\varphi\ (\%) = \frac{(V' - V_{He})}{V'} \times 100 \qquad \text{(Eq. 3.5.2)}$$

The density of the respective monoliths was calculated based on their mass (M') and the volume measured by He-pycnometry (Eq. 3.5.3), thus excluding the open porosity content.

$$\rho = \frac{M'}{V_{He}} \qquad \text{(Eq. 3.5.3)}$$

3.5.3 *Oxidation behavior of uncoated Si_3N_4 substrates*

During pyrolysis and densification of the Si/Yb(2:1) coating system in air, the oxidation of the Si_3N_4 substrate influenced the silicate phase formation and the resulting coating microstructure. More information on the conversion behavior of the Yb(2:1) top-coat was acquired by carrying out oxidation tests with uncoated Si_3N_4 substrates up to the pyrolysis temperature of 1415 °C.

Based on reports about the kinetics of oxidation of Si_3N_4 containing $Y_2O_3/MgO/Al_2O_3$ as sintering additives [73,74,121], higher oxidation rates can be expected at temperatures above 1200 °C due to the formation of a low viscous SiO_2 passivating layer. Thus, uncoated Si_3N_4 substrates were annealed at 750, 1100, 1200 and 1415 °C for 1 h in air in a HT04/18 chamber furnace (Nabertherm GmbH, Germany). A direct comparison with the pyrolysis temperature of the Si/Yb(2:1) coating system was obtained by annealing an additional uncoated Si_3N_4 substrate at 1415 °C for 5 h in air. The long-term oxidation behavior of the selected Si_3N_4 substrates was additionally evaluated by exposing an uncoated Si_3N_4 substrate in a N41/H chamber furnace (Nabertherm GmbH, Germany) at 1200 °C for 200 h in air. The resulting microstructure and phase composition after each oxidation test was analyzed by SEM/EDS and XRD as described in detail in Section 3.5.2.

3.5.4 *Properties of the Si/Yb(2:1) coating system on Si_3N_4*

3.5.4.1 *Phase composition and microstructure*

The microstructure and composition of the deposited Si/Yb(2:1) coatings on Si_3N_4 after pyrolysis at 1415 °C for 5 h in air was analyzed by SEM/EDS as described in detail in Section 3.5.2. Information on the open porosity content was obtained by converting the respective SEM micrographs of their surface into a binary image and quantifying the porosity using the software ImageJ [122]. The crystalline phases were analyzed by XRD measurements at the surface of the coated samples.

3.5.4.2 Adhesion

The adhesion strength of the coating system on Si_3N_4 was measured by pull-off tests using a PosiTest AT-A (DeFelsko Corporation, USA), according to the ASTM D4541. Therefore, ∅10 mm dollies were glued onto the surface of the coated samples with a two-component epoxy resin Loctite EA 9466™ (Henkel Central Eastern Europe GmbH, Austria). The resin was cured at 110 °C for 1 h in air, before the dollies were pulled perpendicularly with a constant force of 1 MPa s^{-1} and the tensile strength necessary to detach them was recorded. The values reported in this work are an average of three measurements. The failure mechanism was examined by SEM/EDS analysis.

3.5.4.3 Hardness

Microhardness measurements were performed by using a Fischerscope H100 (Helmut Fischer GmbH, Germany) at an applied maximum load of 981 mN (HV 0.1) at the surface of the coating, achieving a maximum depth corresponding to 1/10 of the coating thickness, according to the DIN EN ISO 14577.

3.5.4.4 Scratch tests

Scratch tests determined the load necessary to damage the surface of the coatings, which were performed using a Lineartester 249 (Erichsen GmbH & Co. KG, Germany) with varying load ranging from 5 to 40 N. Therefore, a testing rod 15/570 (Ø 1.0 mm) was used, according to the standard ISO 1518-1 and the induced scratches on the coatings were examined by optical microscopy in an Axiotech HAL 100 (Carl Zeiss AG, Germany).

3.5.4.5 Flexural strength

In order to investigate the influence of the coating system on the flexural strength of the Si_3N_4 substrates, 4-point-bending tests were carried out by FCT Ingenieurkeramik at room temperature according to the standard DIN EN 843-1. For this purpose, Si_3N_4 samples with dimensions of 45x4x3 mm were completely coated following the procedure described in Section 3.4. The flexural strength of the coated Si_3N_4 samples was compared with uncoated Si_3N_4 substrates. A total of 14 coated and uncoated samples were respectively tested. The standard deviation for each set of samples was calculated to obtain a representative mean value, and the respective Weibull modulus was determined according to the norm DIN EN 843-5, using the two parameter Weibull distribution function (Eq 3.5.4).

$$P = 1 - \exp\left[\left(-\frac{\sigma}{\sigma_0}\right)^m\right] \qquad \text{(Eq. 3.5.4)}$$

Where "P" is the probability of failure, "m" is the Weibull modulus, "σ" is the flexural strength and "σ_0" the characteristic strength. The Weibull modulus "*m*" was then obtained by transforming Eq. 3.5.4 in Eq. 3.5.5 and plotting "ln ln (1/(1-P))" vs. "ln σ".

$$\ln\ln\left(\frac{1}{1/P}\right) = m\ln(\sigma) - m\ln(\sigma_0) \qquad \text{(Eq. 3.5.5)}$$

The cross-section of the fracture surface of the coated Si_3N_4 was evaluated by optical microscopy in an Axiotech HAL 100 (Carl Zeiss AG, Germany) to determine the failure mechanism.

3.5.4.6 Thermal cycling

Thermal cycling tests evaluated the thermomechanical compatibility between the Si/Yb(2:1) coating system and the Si_3N_4 substrate. Hence, a chamber furnace N41/H (Nabertherm GmbH, Germany) was heated up to 1200 °C and held isothermally at this temperature. In the next step, the uncoated and coated specimens were placed into it and allowed to anneal for 1 h, before quenching in a water bath at 20 °C. A total of 15 cycles were performed.

3.5.4.7 Hot gas corrosion

The potential of the developed coating system for application in gas turbines was assessed by technical hot gas corrosion tests, carried out at Fraunhofer IKTS (Dresden, Germany). For this purpose, 35x4x3 mm Si_3N_4 substrates were completely coated according to the procedure described in Section 3.4. The coated samples were placed inside an Al_2O_3 tube furnace, followed by hot gas corrosion at 1200 °C for 200 h in moist air atmosphere (p = 1 atm, p_{H2O} = 0.15 atm and v = 100 m s^{-1}). After 125 h, the p_{H2O} was reduced from 0.15 to 0.08 atm. The mass change was recorded after pyrolysis (t = 0) and at time intervals of 25, 75, 125 and 200 h by allowing the furnace to cool down to room temperature and weighting the samples, before progressing the test. The mass change reported in this work is an average value obtained from a pair of identical samples. As reference, uncoated Si_3N_4 substrates were likewise corroded. Fig. 3.5.1 shows a scheme of the hot gas corrosion test.

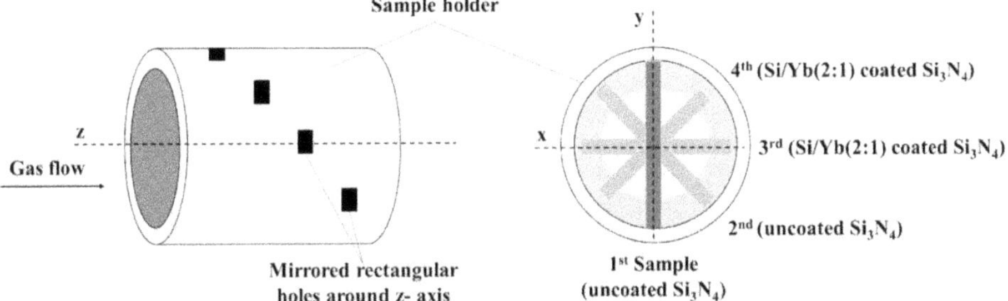

Fig. 3.5.1. Side and front view of the arrangement of the samples during hot gas corrosion tests ($T = 1200$ °C, $v = 100$ m s^{-1}, $t = 200$ h, $p = 1$ atm).

3.5.5 Phase composition and microstructure of the Si/Yb(2:1) coating system on SiC/SiC

The microstructure and composition of the deposited Si/Yb(2:1) coatings on SiC/SiC after pyrolysis at 1415 °C for 5 h in air was analyzed by SEM/EDS and XRD and the open porosity content was determined with help of the software ImageJ as described in detail in Section 3.5.4.1.

3.6 Résumé

Fig. 3.6.1 exhibits a flowchart summarizing the experimental procedures of Section 3.

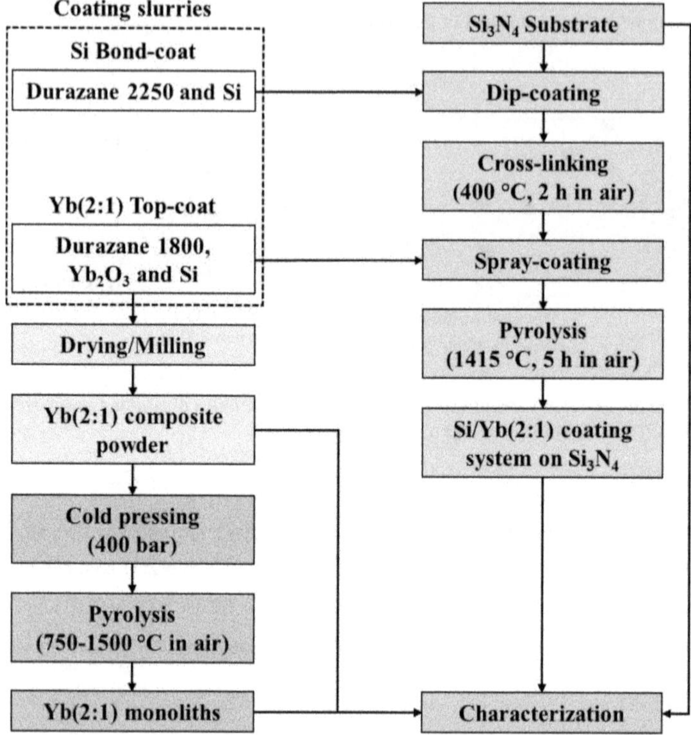

Fig. 3.6.1. Flowchart of the experimental procedures for processing of the Yb(2:1) composite powder and monoliths and the Si/Yb(2:1) coating system on Si_3N_4 by a single deposition step.

In order to understand the pyrolysis behavior of the Yb(2:1) top-coat during the thermal treatment applied for the synthesis and sintering of the $Yb_2Si_2O_7$ phase any influence of the substrate was excluded by investigating powder and monolithic samples processed from the developed Yb(2:1) slurry. Therefore, the slurry was dried under reduced pressure and milled into a fine powder, which was characterized by TGA. Monoliths were then processed by uniaxially pressing the Yb(2:1) composite powder, which were pyrolyzed in the temperature range from 750 to 1500 °C in air. The microstructure and phase composition of the resulting Yb(2:1) monoliths were characterized by SEM/EDS and XRD analyses. The shrinkage and sintering behavior were assessed by comparing the apparent volume of the pressed monoliths before and after pyrolysis. In addition, He-pycnometry measurements determined their true volume, enabling the estimation of the open porosity content and density. Thereby, the ideal temperature necessary to yield a dense and crystalline $Yb_2Si_2O_7$ phase was determined.

First Yb(2:1) coatings on Si_3N_4 revealed the limited oxidation stability of the selected substrates at temperatures above 1400 °C in air. The high temperatures necessary to yield a dense $Yb_2Si_2O_7$ coating led to the enhanced oxidation of the Si_3N_4 substrate and consequently to the formation of blisters, which compromised the mechanical stability of the deposited coatings. In order to understand the influence of the selected Si_3N_4 substrate on the conversion and densification of the Yb(2:1) coating, oxidation tests were carried out with uncoated Si_3N_4 substrates up to 1415 °C in air. The effects of oxidation on the microstructure and phase composition of the substrate were evaluated by SEM/EDS and XRD analyses. The development and application of a Si bond-coat based on Durazane 2250 and elemental Si powder prevented the enhanced oxidation of the Si_3N_4 substrate during pyrolysis in air and hindered the formation of blisters. The double layered coating system was deposited by dipping the Si_3N_4 substrate in the bond-coat slurry, followed by a thermal treatment at 400 °C to yield a stable thermoset coating. Thereafter, the Yb(2:1) slurry was sprayed onto the bond-coated Si_3N_4 substrates, followed by pyrolysis at 1415 °C for 5 h in air.

The microstructure and composition of the resulting Si/Yb(2:1) coating system on Si_3N_4 were characterized by XRD and SEM/EDS analyses. The mechanical properties of the coating were evaluated by pull-off adhesion and scratch tests as well as hardness and 4-point-bending strength measurements. The potential of the developed coating system for protection of non-oxide silicon-based ceramics was assessed by thermal cycling between 1200 °C and quenching in water at 20 °C, as well as hot gas corrosion at 1200 °C for 200 h ($v = 100\ m\ s^{-1}$, $p = 1$ atm and $p_{H2O} = 0.15$ atm), simulating the harsh environments in gas turbines. At last, additional work was carried out on the deposition of thicker Yb(2:1) coatings and on the coating of SiC/SiC.

4 Results and discussion

4.1 Pyrolysis behavior of the Yb(2:1) composite powder

In order to understand the behavior of the Yb(2:1) top-coat during the thermal treatment applied for the synthesis and sintering of the $Yb_2Si_2O_7$ phase, thermogravimetric analysis was performed up to 1415 °C at 5 K min^{-1} and 5 h in air, according to the procedure described in Section 3.5.1. The mass change as a function of the temperature of the Yb(2:1) composite material and the individual components of Table 3.2.2 is exhibited in Fig. 4.1.1.

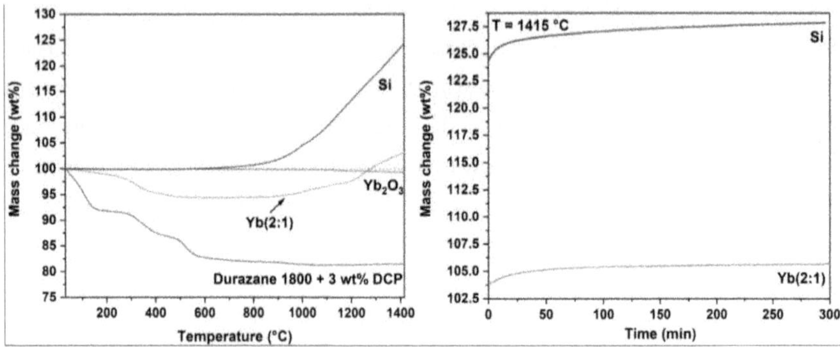

Fig. 4.1.1. (a) Thermogravimetric analysis of the individual components and Yb(2:1) composite material up to 1415 °C at 5 K min^{-1} and (b) 5 h annealing time in synthetic air (80/20 vol% N_2/O_2, 50 ml min^{-1}).

As expected, TG analysis confirms the stability of ytterbium oxide up to 1415 °C in air atmosphere. In contrast, the oxidation of the elemental Si particles starts at approximately 750 °C in air by the formation of an amorphous and dense SiO_2-scale, which limits the inward diffusion of oxygen [71,72,123–127]. The faster diffusion of oxygen through this scale is reported with increasing temperature, accounting for higher oxidation rates, whereas a parabolic and slower rate is exhibited at constant temperatures [71]. This explains the rapid mass gain of 25 wt% detected up to 1415 °C and the final yield of approximately 128 wt% after annealing at 1415 °C for 5 h in air, corresponding to a conversion of 25 mol% of Si into SiO_2.

The oligosilazane Durazane 1800 undergoes a total mass loss of about 19 wt% up to approximately 750 °C, remaining stable up to 1415 °C. According to the literature [47,96,97], the mass loss up to 200 °C can be attributed to the volatilization of low molecular weight oligomers, whereas NH_3, H_2 and their respective oxidation products are eliminated up to 400 °C. The incorporation of oxygen up to this temperature induces the formation of Si-O-Si bonds, leading to a higher degree of cross-linking and increasing the ceramic yield. At above 400 °C, the elimination of organic groups is responsible for the further mass loss up to approximately 750 °C. Nevertheless, the formation of a passivating SiO_2-layer hinders the diffusion of oxygen and the complete elimination of C and N. Hence, a predominantly amorphous ceramic with an approximate molar composition of $SiC_{1.15}N_{0.69}O_{0.41}$ is obtained after pyrolysis at 1600 °C in air [47].

Unlike detected for Durazane 1800 + 3 wt% DCP, the decomposition of Yb(2:1) starts at about 100 °C, which is possibly a consequence of the drying step carried out at 80 °C for 5 h under reduced pressure, leading to the volatilization of oligomers before the TG measurement. Afterwards, the resulting mass loss of 6 wt% up to 750 °C is attributed to the further pyrolysis of the silazane as described above. Beyond this temperature, the oxidation of the Si powder starts and the formation of SiO_2 is responsible for the mass gain, achieving a yield of 103 wt% at 1415 °C. Further annealing at this temperature for another 5 h exhibited a similar trend as measured for elemental Si under the same conditions, denoted by a parabolic oxidation rate. Hence, only a slight mass gain of 2.5 wt% was measured, resulting in a maximum yield of approximately 105 wt%. By considering the composition provided in Table 3.2.2, the final mass changes of the individual components and the complete elimination of the dispersing agent, a theoretical yield of 97 wt% should be achieved. Nevertheless, the complete oxidation of Durazane 1800 and Si into SiO_2 leads to an estimated yield of 107 wt%, which is in good agreement with the measurement of Fig. 4.1.1. Hence, the TG measurement provides evidence for a higher reactivity of Durazane 1800 and Si with oxygen from the atmosphere in the presence of Yb_2O_3.

4.2 Phase composition and microstructure of the Yb(2:1) monoliths

The analysis of the microstructure and composition of the Yb(2:1) monoliths after pyrolysis allowed the study of the formation of the $Yb_2Si_2O_7$ phase and its sintering behavior without the influence of the Si_3N_4 substrate, whereby the ideal temperature for the conversion and densification of the Si/Yb(2:1) coating system was determined. Due to the similar phase composition and microstructure of the monoliths pyrolyzed between 750 and 900 °C, 1200 and 1300 °C as well as 1415 and 1500 °C, only the temperature ranges from 900 to 1100 °C and from 1300 to 1415 °C will be thoroughly discussed. Fig. 4.2.1 shows the respective XRD diffractograms.

At 900 °C, the detection of only crystalline Yb_2O_3 and Si indicates no reaction between the components to yield ytterbium silicate phases. Nevertheless, at 1000 °C the crystalline peaks associated with elemental Si disappeared. This is likely a consequence of its oxidation, supported by the detection of α-cristobalite as well as the incipient formation of ytterbium silicates, thus reducing its fraction. Despite the considerable amounts of Yb_2O_3, the solid-state reaction of Yb_2O_3 with forming SiO_2 yielded the metastable monoclinic P21/c Yb_2SiO_5 and the β-$Yb_2Si_2O_7$ phases. In addition, peaks corresponding to secondary phases as ytterbium oxy-nitride silicates, carbides and silicides were identified, which result from reactions of the pyrolysis products of the oligosilazane and elemental Si with Yb_2O_3. However, the broadness of the detected signals hinders the accurate assignment of these phases, which also indicates a low degree of crystallinity.

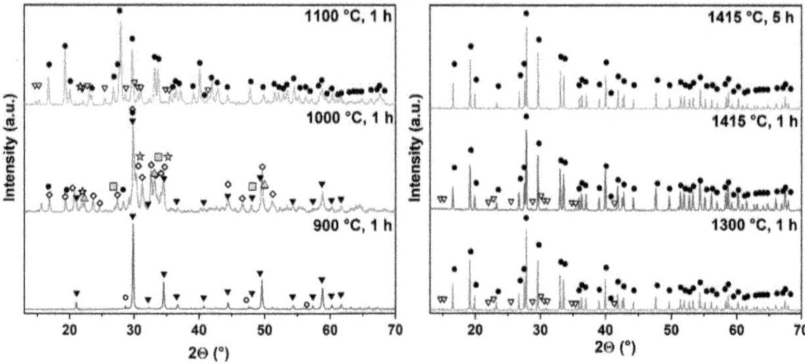

Fig. 4.2.1. XRD analysis of the Yb(2:1) pressed monoliths after pyrolysis in air at different temperatures and annealing times (▼ – Yb_2O_3 (04-001-2438); O – Si (00-027-1402); ☆ α-cristobalite (00-039-1425) ◇- P21/c-Yb_2SiO_5 (00-052-1187); ▽ – I2/a-Yb_2SiO_5 (00-040-0386); ● – β-$Yb_2Si_2O_7$ (00-025-1345); △-$Yb_6Si_{11}N_{20}O$ (04-010-8242); □ - Yb_4Si_7 (04-013-6752); ☆ – Yb_3C_4 (04-003-2283)).

After annealing at 1100 °C for 1 h, the peaks related to the secondary phases disappeared due their low stability in air at high temperatures, forming the respective oxide and silicates [54,128]. The further consumption of Yb_2O_3 leads to the formation of β-$Yb_2Si_2O_7$ and Yb_2SiO_5, which are already main crystalline phases. At this temperature, the transition of the metastable P21/c Yb_2SiO_5 into the equilibrium monoclinic I2/a Yb_2SiO_5 phase was also confirmed, which is in agreement with reported results [129,130]. At higher temperatures, XRD analysis evidences the slow conversion of the monosilicate phase into the disilicate. Hence, only traces of Yb_2SiO_5 are still detected after pyrolysis at 1300 °C, which remain up to 1415 °C. Nevertheless, by annealing the monolith at 1415 °C for another 5 h, XRD analysis indicated the stoichiometric conversion of Yb_2O_3, Si and Durazane 1800 into β-$Yb_2Si_2O_7$. The following SEM/EDS analysis of the corresponding monoliths reveals additional information about their conversion into β-$Yb_2Si_2O_7$ and the sintering behavior of this phase in dependency on the temperature (Fig. 4.2.2).

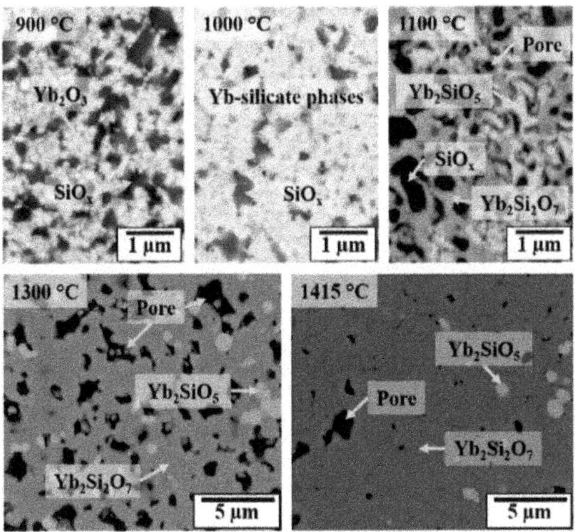

Fig. 4.2.2. SEM micrographs of the cross-section of the Yb(2:1) monoliths after pyrolysis at 900, 1000, 1100 and 1300 °C for 1 h and at 1415 °C for 5 h in air.

After pyrolysis at 900 °C for 1 h, the visible white and black regions correspond to the used Yb_2O_3 and Si particles, respectively. In comparison with Fig. 3.1.2 (b), the use of ZrO_2 milling beads during the slurry preparation (see Section 3.2.2) reduced the mean particle size of the elemental Si powder from approximately 3 µm to less than 1 µm. EDS point analysis of the Si particles detected a non-stoichiometric amount of O regarding the elemental composition of SiO_2, which indicates an initial stage of oxidation in accordance with TG and XRD measurements. These particles were marked as SiO_x. At 1000 °C, the fraction of SiO_x and Yb_2O_3 decreases substantially, which can be directly correlated with the formation of ytterbium silicate phases, denoted by the light grey regions, also confirmed by EDS and XRD analyses. Due to the difficult detection of light elements as carbon and nitrogen by SEM/EDS analysis, the presence of secondary phases such as ytterbium carbide, silicide or nitrogen silicate phases could not be confirmed after pyrolysis at 1000 °C, as otherwise indicated by XRD.

At 1100 °C, ytterbium mono- and disilicate are already well-defined phases, whereby the $Yb_2Si_2O_7$ phase was predominantly detected near the SiO_x particles and is characterized by a compact microstructure. According to XRD and TG analyses, the formation of $Yb_2Si_2O_7$ is almost concluded at 1300 °C, hence only remaining traces of Yb_2SiO_5 were detected by

SEM/EDS, which are embedded in an $Yb_2Si_2O_7$ matrix. Unlike expected, small amounts of Yb_2SiO_5 were still detected after pyrolysis at 1415 °C for 5 h. In this case, the very low relative concentration of this phase in comparison with $Yb_2Si_2O_7$ could have prevented its detection by XRD.

In order to better understand the sintering behavior of the Yb(2:1) monoliths, the volume change ($\Delta V/V_o$) was calculated based on the apparent volume before and after pyrolysis, according to the procedure described in Section 3.5.2.1. The corresponding results for the temperature range of 750 to 1500 °C are exhibited in Fig. 4.2.3.

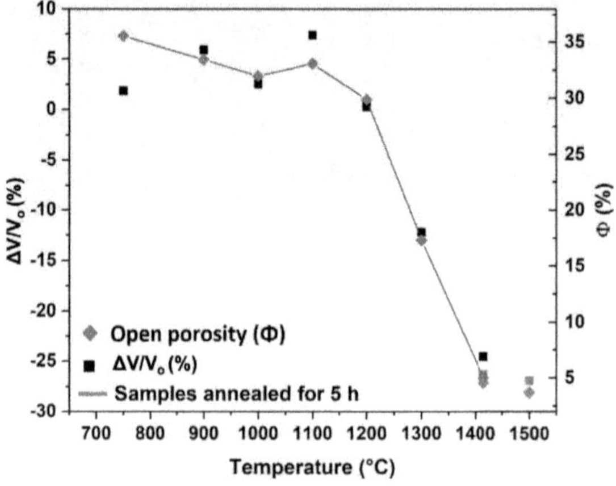

Fig. 4.2.3. Volume change ($\Delta V/V_o$, ■) and open porosity content (Φ, ◆) of the Yb(2:1) pressed monoliths after pyrolysis in air at different temperatures and annealing times.

At 750 °C, a volume change of 2% was measured, which demonstrates that the monoliths retained their apparent volume up to this temperature due to the use of Yb_2O_3 and Si fillers [48]. Besides hindering the shrinkage associated with the pyrolysis and densification of Durazane 1800, this result is also attributed to the partial oxidation of the Si particles, which is accompanied by a molar volume expansion by a factor of 2.2 [71,72,123–127]. As confirmed by TG analysis (refer to Section 4.1), an increasing mass gain was measured for the elemental Si powder in dependency on the temperature, resulting in an apparent volume expansion of 6 vol% at 900 °C. Nevertheless, pyrolysis at 1000 °C led to a slight retraction of 3% in the apparent volume of the pressed monoliths. This is explained by the formation of Yb_2SiO_5 (7.28 g cm^{-3} [31]) and $Yb_2Si_2O_7$ (6.15 g cm^{-3} [131]) from Yb_2O_3 (9.17 g cm^{-3} [31]) and SiO_2 (α-cristobalite, 2.33 g cm^{-3} [61]), leading to a volumetric shrinkage of respectively 9.25 and 11.5%, which was calculated considering the stoichiometric volume of Yb_2O_3 and α-cristobalite necessary to yield ytterbium mono- or disilicate.

The subsequent apparent volume expansion from 3 to 8% within the temperature range of 1000-1100 °C matches the phase transition from P21/c-Yb_2SiO_5 to I2/a-Yb_2SiO_5 and the oxidation of the secondary phases as confirmed by XRD (Fig. 4.2.1), which are accompanied by a large volume expansion [54,129,130]. In addition, the further oxidation of residual Si is expected to lead to an additional volume expansion. At above 1100 °C, the rapid densification of the Yb(2:1) monoliths is evident. Consequently, pyrolysis at 1200 °C resulted in a volume change of 0.3%, which is explained by the conversion of Yb_2SiO_5 into $Yb_2Si_2O_7$ as discussed previously. According to SEM analysis (Fig. 4.2.2), the sintering of the $Yb_2Si_2O_7$ phase leads to a further volume shrinkage of 12% at 1300 °C and 25% at 1415 °C. After annealing the Yb(2:1) monolith at 1415 °C for 5 h, a volume shrinkage of 26% was achieved, which reached a maximum value of 27% after pyrolysis at 1500 °C for 5 h.

He-pycnometry measurements provided additional information about the densification behavior of the Yb(2:1) monoliths after pyrolysis and on the evolution of the open porosity content (Φ) (Fig. 4.2.3). Before pyrolysis, an open porosity of 22 vol% was measured for the Yb(2:1) monoliths, which increased to 36 vol% after pyrolysis at 750 °C. This is primarily a consequence of the shrinkage of the Durazane 1800 during the polymer to ceramic transition [48], as confirmed by TG analysis (refer to Section 4.1). Nevertheless, the apparent volume of the pressed monoliths remained almost constant due to the use of Yb_2O_3 and Si filler particles as already discussed for the volume change, also available in Fig. 4.2.3. Pyrolysis at 900 °C led to a slight decrease in the open porosity content to 33 vol%, which is likely a consequence of the further oxidation of the elemental Si powder. At this temperature, the corresponding SEM micrograph (Fig. 4.2.2) shows no visible pores, which are homogeneously distributed and too small for the used resolution.

Despite the formation of ytterbium silicate and secondary phases at above 1000 °C, no significant change in the porosity content was detected up to 1100 °C. In this case, the constant volume of open porosity may be a consequence of reactions leading to a concurrent volume expansion and shrinkage as evidenced by XRD (Fig. 4.2.1) and SEM/EDS analyses (Fig. 4.2.2). Anyhow, the conversion of Yb_2O_3 and Yb_2SiO_5 into $Yb_2Si_2O_7$ is accompanied by a volume shrinkage, which led to the densification of the Yb(2:1) monolith at 1200 °C as discussed for the volume change, resulting in an open porosity content of 29 vol%. The sintering of the $Yb_2Si_2O_7$ phase at 1300 °C caused a further reduction in the open porosity to 17 vol%, supported by the respective SEM micrograph (Fig. 4.2.2), which shows a dense microstructure and pores with a mean size of 1 μm.

The thermal treatment of the monoliths at 1415 °C for 1 h resulted in a remarkable densification due to the strongly reduced open porosity content of only 5 vol%. Further annealing at 1415 °C for 5 h reduced the open porosity content to 4.5 vol% and only scarce pores can be seen in Fig. 4.2.2. Without considering the open porosity, the monolith had a density of 5.95 g cm^{-3}, which corresponded to 96% of its density after milling (6.19 g cm^{-3}) and is comparable to the bulk density of the β-$Yb_2Si_2O_7$ phase (6.15 g cm^{-3}) [131]). Based on this result, a closed porosity of 4 vol% can be calculated for the monolith, adding up to 8.5 vol% total porosity. This value is considerably lower than reported by Wang *et al.* for the synthesis and sintering of β-$Yb_2Si_2O_7$ monoliths [111], which is among the lowest values reported for the synthesis and pressureless sintering of rare-earth silicates (refer to Section 2.7). In their work, the authors prepared a sol-gel containing Yb_2O_3 powder and tetraethoxysilane (TEOS) in a mol ratio of 2 (TEOS:Yb_2O_3), which was dried at 200 °C for 2 h and pressed at 300 bar. The pressed monolith was pyrolyzed at 1550 °C for 4 h in air, yielding single phase β-$Yb_2Si_2O_7$, which was subsequently milled and screened through a 200-mesh sieve. The resulting β-$Yb_2Si_2O_7$ powder was further cold pressed at 2400 bar and sintered between 1300 and 1550 °C for 4 h in air. After sintering at 1400 °C for 4 h in air, an open porosity of 29.5 vol% was obtained, whereas sintering at 1500 °C led to an open porosity of 11.3 vol% and a density of 5.25 g cm^{-3}.

In the present work, a thermal treatment at 1500 °C for 5 h did not result in significant microstructural changes in comparison with pyrolysis at 1415 °C. Due to the low residual amount of the Yb_2SiO_5 phase and the strongly reduced porosity, pyrolysis at 1415 °C for 5 h in air was selected for the conversion and densification of the Si/Yb(2:1) coating system. Table 4.2.1 gives an overview on the volume change and the open porosity content as well as the phase evolution of the Yb(2:1) pressed monoliths after pyrolysis for different temperatures.

Table 4.2.1. Volume change, open porosity content and phase formation of the Yb(2:1) monoliths after pyrolysis in air for different temperatures and annealing times.

Temperature	Volume change ($\Delta V/V_0$, %)	Open porosity (Φ, vol%)	Phases after pyrolysis
900 °C, 1 h	6	33	Yb_2O_3 and Si
1000 °C, 1 h	3	32	Yb_2O_3, Si, Yb_2SiO_5 (P21/c), β-$Yb_2Si_2O_7$, Yb_3C_4*, $Yb_6Si_{11}N_{20}O$* and Yb_4Si_7*

1100 °C, 1h	8	33	SiO_2, Yb_2O_3, Yb_2SiO_5 (I2/a) and β-$Yb_2Si_2O_7$
1300 °C, 1 h	-12	17	Yb_2SiO_5 and β-$Yb_2Si_2O_7$
1415 °C, 1 h	-25	5	Yb_2SiO_5 and β-$Yb_2Si_2O_7$
1415 °C, 5 h	-26	4.5	Yb_2SiO_5 and β-$Yb_2Si_2O_7$

* detected by XRD, but not confirmed by SEM/EDS.

4.3 Oxidation behavior of uncoated Si_3N_4 substrates

First $Yb_2Si_2O_7$ coatings on Si_3N_4 revealed the limited oxidation stability of the selected substrates at above 1400 °C in air, which contain Al_2O_3, MgO and Y_2O_3 as sintering additives. As shown in Section 4.2, the full densification of the generated $Yb_2Si_2O_7$ phase from the Yb(2:1) composite only occurs after pyrolysis at 1415 °C for 5 h. Up to this temperature and annealing time, oxygen diffused through the porous microstructure of the Yb(2:1) coating and oxidized the Si_3N_4 substrate, which led to the formation of blisters and compromised the mechanical stability of the deposited coatings.

The development and application of a Si bond-coat effectively hindered this drawback and consequently the formation of blisters. Nevertheless, the formation of a passivating SiO_2 layer at the interface between the bond-coat/Si_3N_4 substrate during pyrolysis resulted in the diffusion of sintering additives into it, reducing its viscosity. Consequently, the formed glassy SiO_2 phase infiltrated the Si/Yb(2:1) coating system and accumulated at the grain boundaries of the $Yb_2Si_2O_7$ phase, as will be discussed in Section 4.4.1. In order to understand the influence of the selected Si_3N_4 substrates on the conversion and densification behavior of the Yb(2:1) top-coat, uncoated Si_3N_4 substrates were oxidized at 750, 1100, 1200 and 1415 °C for 1 h and at 1415 °C for 5 h in air, according to the procedure described in Section 3.5.3. The corresponding XRD diffractograms of the oxidized substrates are exhibited in Fig. 4.3.1.

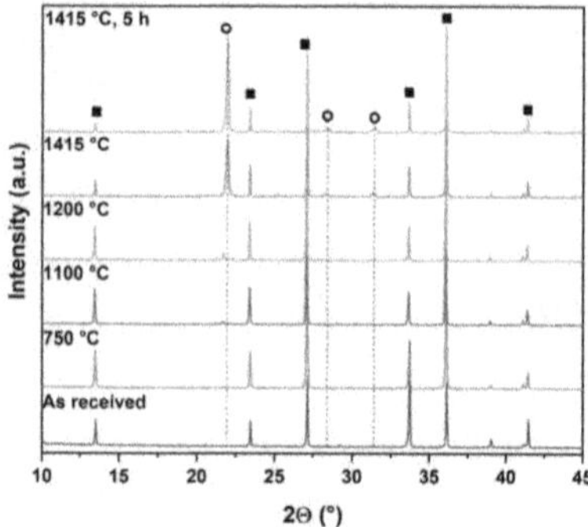

Fig. 4.3.1. XRD analysis of uncoated Si_3N_4 substrates oxidized at 750, 1100, 1200 and 1415 °C for 1 h and at 1415 °C for 5 h in air (■ – β-Si_3N_4 (00-033-1160); **O** – α-cristobalite (00-039-1425)).

Similarly to the untreated substrate, only β-Si_3N_4 was detected up to 750 °C, indicating no oxidation. In contrast, the appearance of a small and broad peak corresponding to α-cristobalite confirms the beginning of the oxidation of the uncoated Si_3N_4 substrates at 1100 °C. Up to 1200 °C, the intensity of α-cristobalite remains almost constant. However, it increases drastically at above this temperature upon the reduction of the intensity of the β-Si_3N_4 phase. This suggests the rapid thickening of the oxidation layer at the surface of the Si_3N_4 substrate, as evidenced by SEM/EDS analysis of the cross-section of the respective Si_3N_4 substrates after oxidation (Fig. 4.3.2).

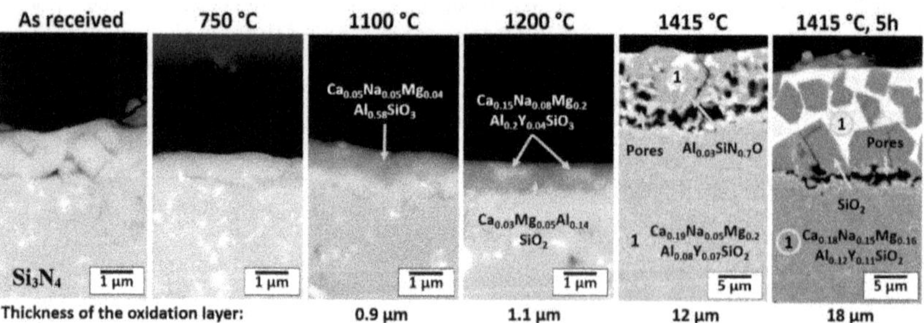

Fig. 4.3.2. SEM micrographs of the cross-section of uncoated Si_3N_4 substrates oxidized at 750, 1100, 1200 and 1415 °C for 1 h and at 1415 °C for 5 h in air.

In agreement with XRD analysis, no signs of oxidation were detected by SEM/EDS analysis up to 750 °C. At 1100 °C, a 0.9 µm thick and well-adherent oxidation layer is formed, composed mainly by Si and O as evidenced by EDS analysis. Nevertheless, trace amounts of Ca, Na, Al and Mg were also detected, according to the composition marked in the corresponding SEM micrograph. This confirms the diffusion of sintering additives and impurities from the bulk of the Si_3N_4 substrate into the oxidation layer. As expected, no significant microstructural changes were detected between 1100 and 1200 °C. At 1200 °C, the oxidation layer achieved a thickness of 1.1 μm and a slightly higher concentration of metal impurities was detected. This resulted in the formation of two SiO_2 phases, which can be distinguished by the presence of Y.

After annealing at 1415 °C for 1 h in air, the phase separation becomes evident. The EDS composition of the white phase (region 1) still resembles that of the oxidation layer at 1200 °C. However, the light grey phase is denoted by a composition of $Al_{0.03}SiN_{0.7}O$ (see Fig. 4.3.2). The detection of small amounts of N indicate the formation of an intermediary oxidation phase between SiO_2 and Si_3N_4, corresponding to Si_2N_2O, as commonly reported during oxidation experiments of Si_3N_4 [15,18,72,73]. The 12 μm thick oxidation layer is denoted by a porous microstructure, likely arising from the entrapment of gases released during oxidation at 1415 °C. Further annealing at 1415 °C for 5 h led to the densification of the oxidation layer, which achieved a total thickness of 18 μm. The oxidation of the $Al_{0.03}SiN_{0.7}O$ phase resulted in the formation of stoichiometric SiO_2, whereas the further diffusion of sintering additives and impurities increased the concentration of Na and Y and the fraction of region 1. The absence of peaks related to region 1 in the respective XRD diffractogram of Fig. 4.3.1 suggests the amorphous character of this phase.

The effects of long-term oxidation on the microstructure and phase composition of uncoated Si_3N_4 substrates were further evaluated after oxidation tests at 1200 °C for 200 h in air. Fig. 4.3.3 shows the respective SEM/EDS and XRD analyses after oxidation.

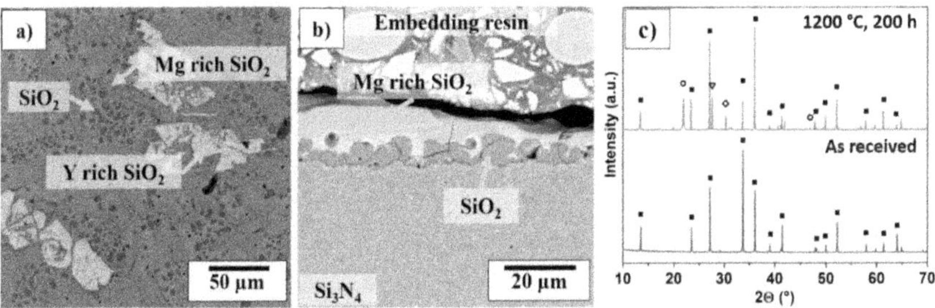

Fig. 4.3.3. SEM micrograph of the (a) surface and (b) cross-section of the uncoated Si_3N_4 substrate after oxidation at 1200 °C for 200 h in air. (c) Corresponding XRD diffractogram before and after oxidation (■ – β-Si_3N_4 (00-033-1160); **O** – α-cristobalite (00-039-1425); ▽ – $Y_2Si_2O_7$ (00-022-1103); ◊ – $MgSiO_3$ (04-020-1659)).

After exposure for 200 h at 1200 °C in air, three distinct phases were detected by EDS spot analysis at the surface of the Si_3N_4 substrate (Fig. 4.3.3 (a)), corresponding to stoichiometric SiO_2 and Y and Mg rich SiO_2. In both metal rich SiO_2 phases, the presence of approximately 18 at% Ca and less than 3 at% Na and Al resembles the composition of the oxidation layer of the Si_3N_4 substrate after exposure at 1200 °C for 1 h in air (Fig. 4.3.2). However, the long annealing time favored the accumulation of metal impurities in the oxidation layer. Based on these results, an empirical composition of $Ca_{0.6}Y_{0.25}SiO_{1.5}$ can be estimated for Y rich SiO_2 and $Ca_{0.6}Mg_{0.47}SiO_{1.6}$ for Mg rich SiO_2. The analysis of the cross-section of the oxidized Si_3N_4 substrate reveals a 15 μm thick oxidation layer, composed predominantly by Mg rich SiO_2, which is present at its outer surface, and stoichiometric SiO_2 as an interphase to Si_3N_4 (Fig. 4.3.3 (b)). XRD analysis confirms β-Si_3N_4 as primary crystalline phase, arising from the unoxidized substrate underneath the oxidation layer (Fig. 4.3.3 (c)). The detection of α-cristobalite is in good agreement with the phase composition evidenced by EDS analysis. In addition, main crystalline peaks of $Y_2Si_2O_7$ and $MgSiO_3$ can be assigned to the respective X-ray diffractogram, although these phases could not be clearly distinguished by EDS analysis.

Due to the difficult sintering of Si_3N_4 ceramics, additives such as Al_2O_3, MgO and rare-earth oxides (RE_2O_3) help densify it by the formation of silicate melts through the reaction with a naturally grown and passivating SiO_2 layer at the surface of the Si_3N_4 feedstock powder. Already during sintering, metal impurities segregate from the bulk of the Si_3N_4 feedstock powder and other raw materials, migrating to the SiO_2 rich grain boundary phase due to a difference in the chemical potential. When oxidation starts at temperatures approximately above 750 °C in air, a dense and amorphous SiO_2 layer is formed at its surface according to the mechanism of Eq. 4.3.1 and 4.3.2 [15,18,72,73].

$$Si_3N_4 + 0.75\ O_2 \rightarrow 1.5\ Si_2N_2O + 0.5\ N_2 \qquad \text{(Eq. 4.3.1)}$$

$$Si_2N_2O + 1.5\ O_2 \rightarrow 2\ SiO_2 + N_2 \qquad \text{(Eq. 4.3.2)}$$

The concentration gradient of metal impurities and additives between the grain boundary phase and the forming SiO_2 at its surface creates a driving force for their diffusion until equilibrium is reached. Since SiO_2 is continuously being formed, the constant gradient leads to the accumulation of metals in the oxidation layer, despite their low concentration within the bulk. This results in the precipitation of silicates and mixed silicate phases at the outer surface of the SiO_2 layer [19,72,73,76,78,118,119,132,133], as evidenced by EDS and XRD analyses (Fig. 4.3.3). Moreover, it induces the crystallization of SiO_2 at above 1000 °C in dry atmospheres [61,71,134–137]. The volume shrinkage associated with it and the subsequent transformation of β → α cristobalite at 270 °C during cooling cause the formation of cracks, as seen in Fig. 4.3.2 (T = 1415 °C) and Fig. 4.3.3 (b).

Generally, the diffusion of cationic species is strongly dependent on the generated field strength between the pair cation and anion, which is proportional to the ion charge and inversely proportional to the ion size [138,139]. Hence, in the pairs CaO and MgO, Ca^{2+} will always diffuse at a faster rate than Mg^{2+}, explaining the higher Ca content within the oxide layer of Fig. 4.3.3. The higher diffusion coefficients of Ca^{2+} and Mg^{2+} in comparison with Y^{3+} lead to their accumulation at lower oxidation temperatures (1000-1300 °C), whereas their oversaturation and the increased mobility of the larger rare-earth cations above 1300 °C decrease this disparity in concentration [78]. Nevertheless, factors as the temperature, Si_3N_4 microstructure, the amount, distribution, composition and crystallinity of the grain boundary phase also influence their diffusion behavior [17,78].

By acting as network modifiers, the cation impurities reduce the viscosity of the passivating SiO_2 layer, thus enabling the easier diffusion of oxygen and enhancing the oxidation of Si_3N_4. Beyond this point, the diffusion of cation impurities becomes the rate-limiting step of oxidation [16,17,19,76,78,118,119,133,138]. Especially Ca^{2+} and Mg^{2+} can greatly reduce the viscosity and the softening point of glasses to 1100-1300 °C [119,140–142], whereby melting points lower than 1350 °C are reported for magnesium and calcium silicates containing Na or Al [143,144]. Finally, the low viscosity of the oxidation layer at 1415 °C could explain the entrapment of gases arising during the oxidation of the Si_3N_4 substrate, forming a porous microstructure after oxidation for 1 h (Fig. 4.3.2), which also provides evidence about the formation of blisters in the first coating experiments.

4.4 Properties of the Si/Yb(2:1) coating system on Si_3N_4

4.4.1 Phase composition and microstructure

After the definition of the pyrolysis parameters based on the formation of the $Yb_2Si_2O_7$ phase and the porosity content as demonstrated in Section 4.2, Si_3N_4 substrates were coated with the Yb(2:1) slurry and pyrolyzed at 1415 °C for 5 h in air. The coating material is very porous up to 1300 °C, which led to the enhanced oxidation of the Si_3N_4 substrate before the densification of the coating, causing the formation of blisters, as discussed in detail in Section 4.3. Therefore, the deposition of a Si bond-coat slurry composed of Durazane 2250 and elemental Si powder prior to the Yb(2:1) coating was necessary to protect the substrate from oxidation. The resulting double-layer Si/Yb(2:1) coating system was pyrolyzed at 1415 °C for 5 h in air.

SEM/EDS analysis of the surface of the coating system on Si_3N_4 after pyrolysis reveals a dense microstructure with an open porosity content of 4 vol% as estimated by image analysis using the software ImageJ (Fig. 4.4.1 (a)). Unlike the Yb(2:1) monoliths (refer to Section 4.2), the shrinkage of the Yb(2:1) coating during pyrolysis may only occur freely in thickness due to the adhesion of the coating system to the substrate. In this case, the concurrent thermal expansion of the Si_3N_4 substrate hinders its densification, which can explain the formation of porosity. At higher magnifications (Fig. 4.4.1 (b)) round shaped grains with up to 5 µm in diameter and a binder phase can be distinguished, also characterized by a different elemental composition.

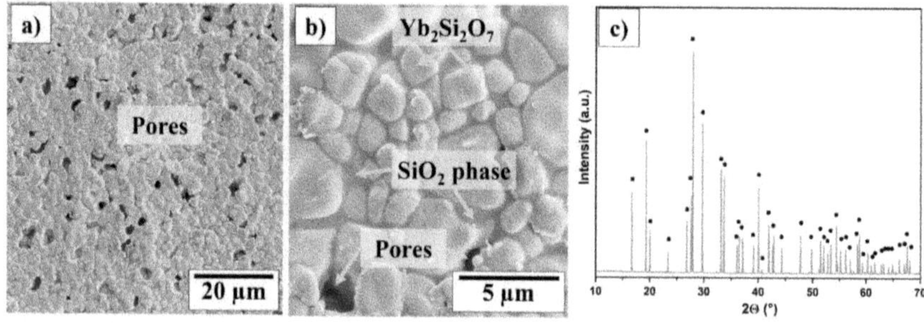

Fig. 4.4.1. (a and b) SEM micrograph of the surface of the Si/Yb(2:1) coating system on Si_3N_4 and (c) XRD analysis after pyrolysis at 1415 °C and 5 h in air (• – β-$Yb_2Si_2O_7$ (00-025-1345)).

The composition of the grain phase is consistent with $Yb_2Si_2O_7$ as 60 at% O, 19 at% Si and 21 at% Yb were detected by EDS spot analysis, whereas the binder phase is composed of SiO_2 (26 at% Si and 62 at% O). However, trace impurities of Ca, Mg, Na and Al were also detected, accounting respectively for less than 4 at%. XRD analysis of the surface of the Si/Yb(2:1) coating system on Si_3N_4 after pyrolysis at 1415 °C for 5 h in air (Fig. 4.4.1 (c)) shows the formation of only β-$Yb_2Si_2O_7$, which is in good agreement with the corresponding Yb(2:1) monolithic sample (refer to Fig. 4.2.1). The absence of crystalline peaks corresponding to SiO_2 in the XRD diffractogram of Fig. 4.4.1 (c) indicates the amorphous character of the binder phase.

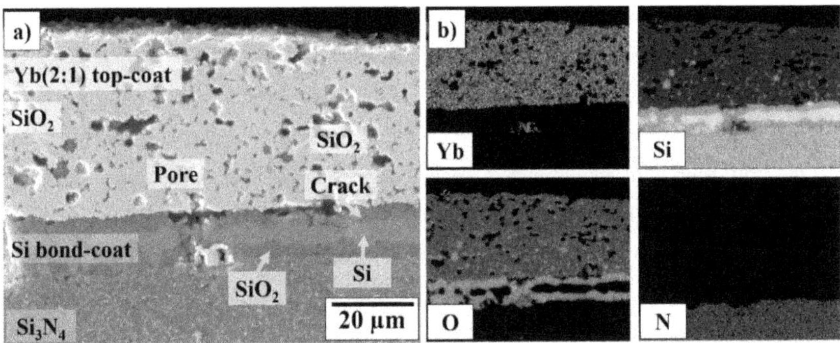

Fig. 4.4.2. (a) SEM micrograph and (b) EDS mapping of elements of the cross-section of the Si/Yb(2:1) coating system on Si_3N_4 after pyrolysis at 1415 °C and 5 h in air.

The SEM analysis of the cross-section of the Si/Yb(2:1) coating system shows well-adherent layers with a thickness of approximately 68 µm (Fig. 4.4.2 (a)). The EDS mappings of elements is exhibited in Fig. 4.4.2 (b) and allows the distinction between the different layers. The thickness of the Yb(2:1) top-coat was estimated in 58 µm, which can be identified by the Yb mapping. The bond-coat is 10 µm thick and is composed of only Si and O. By comparing the Si and O mappings, it is evident that the elemental Si powder melted within the bond-coat during pyrolysis, leading to a dense microstructure, which effectively protected the Si_3N_4 substrate against oxidation and hindered the formation of blisters during the thermal treatment applied for the densification of the top-coat.

The black spots, mainly visible at the surface region of the Yb(2:1) coating (see Si mapping, Fig. 4.4.2 (b)), indicate pores within the coating system. In contrast, the amorphous SiO_2 phase is predominantly present closer to the Si bond-coat and at the grain boundaries of the $Yb_2Si_2O_7$ phase, identified by the higher intensities of Si and O, in agreement with SEM analysis of the surface of the coating (Fig. 4.4.1 (b)). EDS spot analysis of this phase in the cross-section of the Yb(2:1) top-coat also confirmed the presence of stoichiometric SiO_2, formed by the oxidation of the elemental Si particles during pyrolysis in air and marked as SiO_2 in Fig. 4.4.2 (a). SiO_2 was also detected within the bond-coat, especially at the interfaces to the Si_3N_4 substrate and to the Yb(2:1) coating, enclosing elemental Si (compare Si and O mappings). This indicates the oxidation of the bond-coat/Si_3N_4 substrate interface during pyrolysis, which can explain the origin of the amorphous binder phase.

As discussed in Section 4.2, the densification of the coating system only occurs only at 1415 °C. Before a dense microstructure is achieved, the oxidation of the interface between the bond-coat/Si_3N_4 substrate led to the formation of a SiO_2 passivating layer. As already mentioned, the Si_3N_4 substrate contains Y_2O_3, MgO and Al_2O_3 as sintering aids. The concentration gradient between these oxides and metal impurities within the glassy grain boundary phase in the substrate and the formed SiO_2 passivating layer creates a driving force for their diffusion until equilibrium is reached. By acting as network modifiers, the cation impurities reduce the viscosity of the SiO_2 layer [78,118,133], as discussed in detail in Section 4.3. The low viscosity of the generated glassy SiO_2 phase favored the infiltration of the coating system, which resulted in a denser microstructure in the deeper coating regions. Regarding the corresponding Yb(2:1) monolith (Fig. 4.2.1), the excess of glassy SiO_2 contributed to the conversion of residual Yb_2SiO_5 into $Yb_2Si_2O_7$. This explains the presence of unreacted SiO_2 within the Yb(2:1) top-coat in Fig. 4.4.2 (a) and the accumulation of glassy SiO_2 at the grain boundaries of $Yb_2Si_2O_7$. Several authors report that already small amounts of cation impurities induce the crystallization of SiO_2 at above 1000 °C in dry atmospheres [61,71,134–137]. The associated volume shrinkage and the subsequent transformation of $\beta \rightarrow \alpha$ cristobalite at 270 °C during cooling cause the formation of cracks, as evidenced within the bond-coat in Fig. 4.4.2 (a).

4.4.2 Adhesion

Pull-off tests carried out with coated Si_3N_4 substrates according to the ASTM D4541 confirmed an adhesion strength of 36.9 ± 6.2 MPa, higher than reported for other ytterbium disilicate EBCs/Si bond-coated systems on SiC (29.1 ± 0.8 MPa) [145]. The outstanding adhesion was attributed to the good chemical compatibility between the silazane-based Si bond-coat and the Si_3N_4 substrate [22,35,86,146], besides the infiltration with glassy SiO_2, which may have favored the mechanical anchoring of the coating system. The failure mechanism was analyzed in detail by SEM/EDS (Fig. 4.4.3 (a) and (b)), where C and Al mappings represent the resin and the aluminium dolly counterpart used in the pull-off test, respectively.

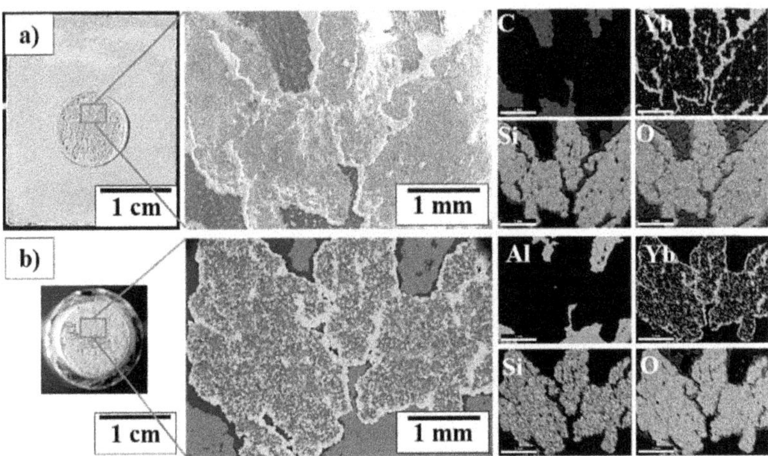

Fig. 4.4.3. Optical and respective SEM/EDS micrographs of the surface of the (a) Si/Yb(2:1) coating system on Si_3N_4 and (b) Al dolly counterpart after pull-off test.

The absence of Yb and the high intensities of Si and O, concurrently detected at the surfaces of the tested sample and the Al dolly, confirm that the bond-coat failed by cohesion. This was likely favored by the crystallization of SiO_2 and the formation of small cracks in the bond-coat (see Fig. 4.4.2 (a)). Nevertheless, the pull-off test also evidences the good compatibility and the excellent adhesion strength between the top and bond-coat layers as well as the strong cohesion of the *in situ* generated $Yb_2Si_2O_7$ crystallites.

4.4.3 Hardness

Microhardness measurements at the surface of coated samples were carried out with a maximum applied load of 981 mN (HV 0.1), reaching a maximum depth corresponding to 1/10 of the coating thickness, according to the DIN EN ISO 14577. Thereby, a hardness of 713 ± 169 HV (6.9 ± 1.6 GPa) was determined, which is comparable to the intrinsic hardness of $Yb_2Si_2O_7$ (7.3 ± 0.2 GPa) [131] and is also an indication for the low residual porosity content of 4 vol% of the coating system (Fig. 4.4.1).

4.4.4 Scratch tests

To complement the mechanical investigations, scratch tests were carried out according to the standard ISO 1518-1, which demonstrated that a critical load of 28 N is necessary to damage the coating system. The damaged surface was evaluated by light microscopy and is exhibited in Fig. 4.4.4.

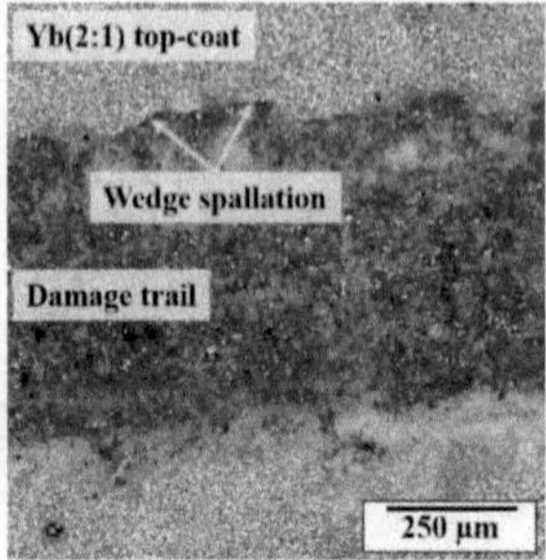

Fig. 4.4.4. Light microscopy analysis of the surface of the Si/Yb(2:1) coating system on Si_3N_4 after scratch tests (applied load: 28 N).

The analysis of the damaged area reveals a damage trail with a width of approximately 625 µm, evidenced by the spallation of the coating and the formation of wedges. This is in good agreement with the typical behavior of hard and well-adhered coatings on hard substrates [147,148], where plastic deformation is minimal and fracture dominates the scratch response. Hence, the good mechanical response of the Si/Yb(2:1) coating system to scratch tests can be attributed to the low porosity content of 4 vol% with only minor defects, high hardness (6.9 ± 1.6 GPa) and adhesion strength (36.9 ± 6.2 MPa) to the Si_3N_4 substrate.

4.4.5 Flexural strength

In order to evaluate the influence of the coating system on the flexural strength of the Si_3N_4 substrate, 4-point-bending tests were carried out by FCT Ingenieurkeramik at room temperature, according to the standard DIN EN 843-1 and the procedure described in Section 3.5.4.5. The mean flexural strength and standard deviation of uncoated Si_3N_4 substrates and Si/Yb(2:1) coated Si_3N_4 are exhibited in Fig. 4.4.5.

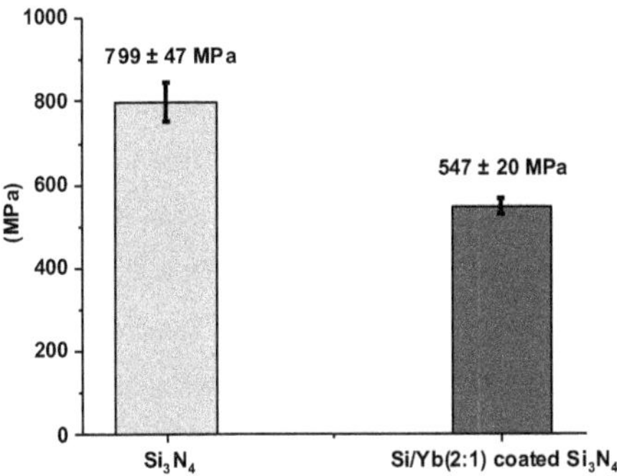

Fig. 4.4.5. Room temperature 4-point bending test of uncoated Si_3N_4 substrates and Si/Yb(2:1) coated Si_3N_4.

From Fig. 4.4.5, a flexural strength of 799 ± 47 MPa was determined for the uncoated Si_3N_4 substrate, which is in good agreement with the data provided by FCT Ingenieurkeramik for SNPU (σ = 870 MPa), considering the standard deviation [117]. After the application of the Si/Yb(2:1) coating system onto Si_3N_4 followed by pyrolysis at 1415 °C for 5 h in air, a flexural strength of 547 ± 20 MPa was measured, which accounts for a debit in strength at room temperature of approximately 32%, regarding the uncoated substrates. The Weibull analysis of both set of samples is shown in Fig. 4.4.6, where the coefficient "*m*" in the equation "y = *m*x + b" corresponds to the Weibull modulus.

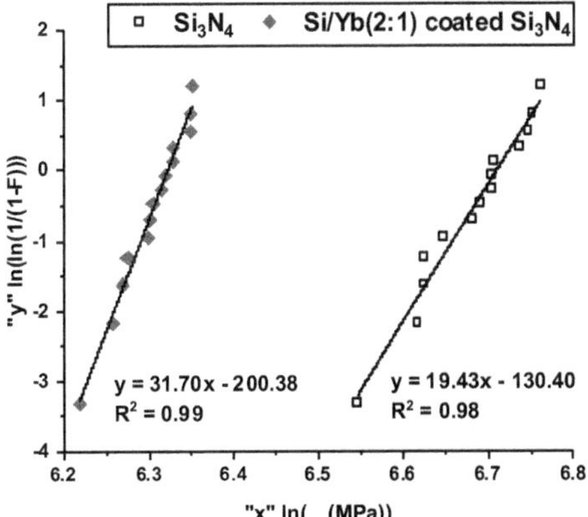

Fig. 4.4.6. Weibull analysis of the room temperature 4-point-bending strength measurements of uncoated Si_3N_4 substrates and Si/Yb(2:1) coated Si_3N_4.

In comparison with the coated Si_3N_4, the flexural strength measurements of the uncoated Si_3N_4 substrate presented a higher variability, which resulted in a Weibull modulus of 19.43. The high Weibull modulus evidences the uniform distribution of defects after the processing of the Si_3N_4 substrates and allows the good prediction of the failure strength [149]. The same statement is valid for the Si/Yb(2:1) coated Si_3N_4 substrates, which presented a narrower data distribution caused by the systematic introduction of defects by the coating system, resulting in a Weibull modulus of 31.70. The optical analysis of the cross-section of the fracture surface of the Si/Yb(2:1) coated Si_3N_4 substrate indicates that the crack initiated at the border and coating region, which is marked by a red circle in Fig. 4.4.7 and is most likely the cause of failure.

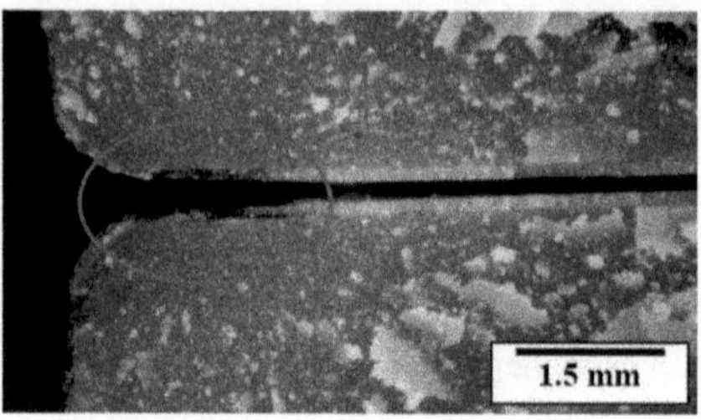

Fig. 4.4.7. Optical micrograph of the cross-section of the fracture surface of the Si/Yb(2:1) coated Si_3N_4 and the corresponding counterpart after 4-point-bending test at room temperature.

The debit in strength at room temperature of coated monolithic Si_3N_4 has been well documented by Bhatia *et al.* [26,36]. In their reports, the authors investigated the flexural strength of SN282 and AS800, commercially available turbine grade Si_3N_4 ceramics, coated by air plasma spray with a Si bond-coat and BSAS top-coat. After the coating deposition, a crystallization step was carried out at 1250 °C for 24 h in air. Despite the small mismatch, the different CTEs of the BSAS top-coat (CTE = 4-5 10^{-6} K^{-1} [22]), the Si bond-coat (CTE = 4.1 10^{-6} K^{-1} [10]) and the Si_3N_4 substrate (CTE = 3.60 10^{-6} K^{-1} [19]) led to the formation of a cracked coating microstructure. In comparison with the uncoated SN282 and AS800 substrates, a debit in strength of up to 60% was reported for the coated substrates at room temperature, decreasing from approximately 790 to 317 MPa.

The analysis of the fracture surface of the coated specimens revealed that the fracture initiated at the coating region, similarly to Fig. 4.4.7. The debit in strength was caused by residual thermal stresses at room temperature due to the CTE mismatch between coating system and substrate, which was intensified by the strong bonding of the coating system and the interaction of its cracks with pre-existing flaws in the Si_3N_4 ceramic substrates. Nevertheless, due to the lower residual thermal stresses at higher temperatures a debit in strength of only 13%

was determined for coated Si_3N_4 substrates by carrying out flexural strength measurements at 1200 °C. In comparison with the system of Bhatia *et al.*, the microstructure of the Si/Yb(2:1) coating system is denoted by a lower flaw density (refer to Fig. 4.4.2), which can explain the higher strength retention at room temperature. Moreover, lower residual thermal stresses are expected due to the lower CTE of the $Yb_2Si_2O_7$ phase (3.91 10^{-6} K^{-1} [19]) within the Yb(2:1) top-coat in comparison with BSAS, thus increasing the stability of the Si/Yb(2:1) coated Si_3N_4 substrate at room and high temperatures.

4.4.6 Thermal cycling

The thermomechanical compatibility between the developed Si/Yb(2:1) coating system and the Si_3N_4 substrate was evaluated by thermal cycling, carried out by annealing coated samples at 1200 °C for 1 h in air, before quenching in a water bath at 20 °C. Optical micrographs of the corresponding coated sample after 5, 10 and 15 cycles are exhibited in Fig. 4.4.8.

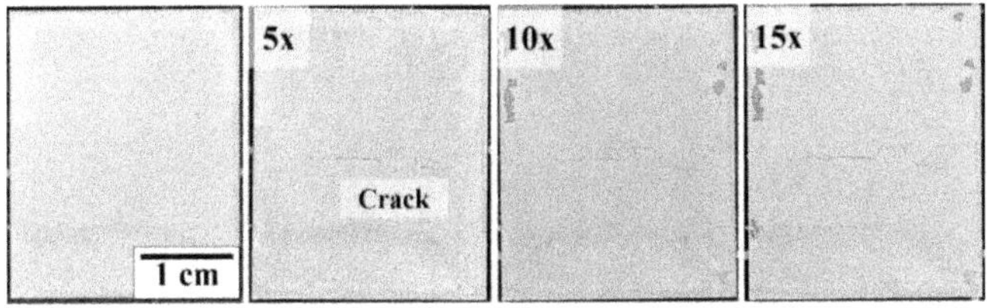

Fig. 4.4.8. Optical micrographs of the surface of the Si/Yb(2:1) coating system on Si_3N_4 after thermal shock cycles between 1200 °C and quenching in a water bath at 20 °C.

By quenching the tested sample in a water bath at 20 °C, the low mechanical stability of the monolithic Si_3N_4 substrate can be expected due to its brittle behavior in combination with the high temperature gradient and the extreme coefficient of heat transfer. For this reason, macroscopic cracks were already formed within substrate after the second cycle, which did not influence the adhesion of the coating system. In this case, it remained completely adhered up to 6 thermal cycles, whereas only few spallation spots occurred at the edges of the coated Si_3N_4 substrate after 15 cycles, indicated by the yellow arrows (Fig. 4.4.8).

Besides the good CTE compatibility between the coating system and the Si_3N_4 substrate (refer to Section 1), the good performance during thermal shock tests was also attributed to the residual porosity within the microstructure of the coating and the softening of the glassy SiO_2 phase at high temperatures. The residual porosity decreases the Young's modulus of the coating, thus increasing its thermal shock tolerance, and the softening of the glassy SiO_2 phase during annealing at 1200 °C likely led to the relaxation of stresses generated upon thermal cycling [150,151]. Despite the extreme thermal shock conditions, the excellent compatibility between the Si/Yb(2:1) coating system and the Si_3N_4 substrate is clear. The thermal cycling fatigue of the selected Si_3N_4 substrates could be better assessed by milder testing conditions.

4.4.7 Hot gas corrosion

Finally, the potential of the developed Si/Yb(2:1) coating system for protection of non-oxide silicon based ceramics against very harsh hot gas corrosion atmospheres was investigated at 1200 °C and a gas flowing speed of 100 m s^{-1} for 200 h, simulating the operating conditions seen by turbine blades and vanes in the hot section of gas turbines [6,15]. The detailed procedure of the hot gas corrosion tests is described in Section 3.5.4.7. For a matter of comparison, uncoated Si_3N_4 substrates were likewise corroded. Fig. 4.4.9 displays the corresponding optical micrographs of the samples before and after hot gas corrosion test.

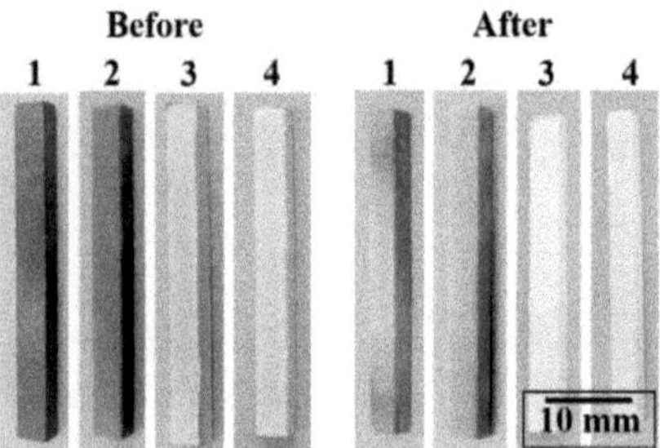

Fig. 4.4.9. Optical micrographs of the uncoated Si_3N_4 substrates (1 and 2) and Si/Yb(2:1) coated Si_3N_4 (3 and 4) before and after hot gas corrosion test at 1200 °C for 200 h (v = 100 m s^{-1}).

After exposure at 1200 °C for 200 h and a gas flow speed of 100 m s^{-1}, the surface of the uncoated samples acquired a white and opaque color, indicating the formation of an oxidation layer (samples 1 and 2). In contrast, the coated samples appeared intact and remained well-adhered to the Si_3N_4 substrate (samples 3 and 4). The specific mass loss over time and XRD analysis of the corroded samples are shown in Fig. 4.4.10.

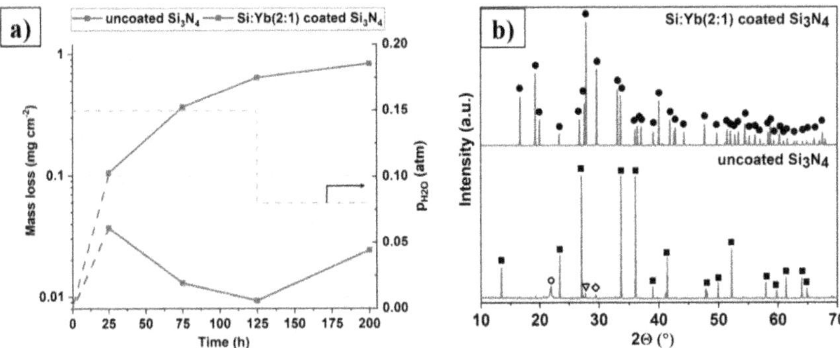

Fig. 4.4.10. (a) Mass loss over time and (b) XRD analysis of the tested samples after hot gas corrosion test at 1200 °C for 200 h (v = 100 m s^{-1}) (■ – β-Si_3N_4 (00-033-1160); **O** – α-cristobalite (00-039-1425); ▽ – $Y_2Si_2O_7$ (00-022-1103); ◊ – $MgSiO_3$ (04-020-1659); • – β-$Yb_2Si_2O_7$ (00-025-1345)).

As expected, the high flow rate led to the rapid degradation of the Si_3N_4 substrates in hot gas environments at 1200 °C, achieving a total mass loss of 0.83 mg cm^{-2} after 200 h. In comparison with uncoated substrates, the coating system reduced mass loss by 98% after 200 h, which corresponded to 0.02 mg cm^{-2}. In combustion environments, corrosion and oxidation occur simultaneously and contribute both to the mass change. In this case, the mass loss measured for both uncoated and coated Si_3N_4 substrates after 200 h indicates the predominance of corrosion reactions. The decrease in the corrosion rate over time is noticeable for the uncoated Si_3N_4 substrate after 125 h, caused by the reduced concentration of water vapor in the atmosphere due to the lower p_{H2O}. Up to 125 h, the corrosion curve of the uncoated Si_3N_4 substrate can be modeled by the equation "$\mathbf{\Delta M}$ = 5.05 10^{-3} **t**" with an R^2 of 0.998, where $\mathbf{\Delta M}$ and **t** correspond to the mass loss (mg cm^{-2}) and time (h), respectively, indicating the linear recession of the sample, in agreement with the work of Opila *et al.* and Fox *et al.* [16–18]. By considering the 200 h of testing, the quality of the linear fit worsens (R^2 = 0.988) and the corrosion rate reduces from 5.05 10^{-3} mg cm^{-2} h^{-1} to 4.49 10^{-3} mg cm^{-2} h^{-1}, due to the lower p_{H2O} as explained before.

For the coated Si_3N_4 substrate, after an initial mass loss because of corrosion, the clear dominance of oxidation reactions after 25 h led to a gradual gain in mass up to 125 h. However, in the following period up to 200 h a low mass loss was detected despite the reduction of the p_{H2O} to 0.08 atm, which will be explained later in detail. Due to the irregular corrosion behavior, the mathematical modelling of the corrosion curve of the Si/Yb(2:1) coated Si_3N_4 substrate is not practical and the discussion of the resulting mass loss is only meaningful by analyzing the microstructure of the corroded sample.

In wet environments, the oxidation rate of Si_3N_4 is up to 20 times higher than in dry O_2 due to the increased permeation of SiO_2 by H_2O (Eq. 4.4.1), which is enhanced by the oxidation of the released NH_3, increasing the fraction of H_2O in the atmosphere (Eq. 4.4.2). Si_3N_4 usually forms a protective and efficient silica layer at high temperatures in air, however at temperatures above 1200 °C, the volatilization of $Si(OH)_4$ leads to a significant mass loss and surface recession of the Si_3N_4 ceramic (Eq. 4.4.3). Moreover, similar effects as discussed for dry oxidation of Si_3N_4 ceramics containing sintering additives (refer to Section 4.3) were reported for wet environments. These include the faster crystallization of SiO_2, the diffusion of sintering additives and impurities from the bulk of Si_3N_4, their precipitation as silicates and the reduction of the viscosity of the forming SiO_2 layer, which increase oxidation and corrosion rates [119,121,152–155].

$$Si_3N_4 + 6\ H_2O \rightarrow 3\ SiO_2 + 4\ NH_3 \qquad \text{(Eq. 4.4.1)}$$

$$4\ NH_3 + 5\ O_2 \rightarrow 6\ H_2O + 4\ NO \qquad \text{(Eq. 4.4.2)}$$

$$SiO_2 + 2\ H_2O \rightarrow Si(OH)_4 \qquad \text{(Eq. 4.4.3)}$$

After the exposure for 200 h at 1200 °C during hot gas corrosion test, XRD analysis of the uncoated Si_3N_4 substrate detected SiO_2 as α-cristobalite and main crystalline peaks of $Y_2Si_2O_7$ and $MgSiO_3$ (Fig. 4.4.10 (b)), similarly to the results obtained after dry oxidation tests of uncoated Si_3N_4 substrates at 1200 °C for 200 h in air (refer to Fig. 4.3.3, Section 4.3). In this regard, SEM/EDS analysis of the surface of the Si_3N_4 substrate also confirmed the formation of stoichiometric SiO_2, besides Y and Mg rich SiO_2 (Fig. 4.4.11 (a)). Based on the EDS results, a composition of $Ca_{0.03}Al_{0.05}Y_{0.03}SiO$ can be estimated for Y rich SiO_2 and $Ca_{0.9}Al_{0.04}Mg_{0.07}SiO_{2.7}$ for Mg rich SiO_2. Anyhow, no clear distinction of $Y_2Si_2O_7$ and $MgSiO_3$ phases was possible by SEM/EDS analysis due to the scarce detection of Y and Mg. Besides the measured mass loss, the degradation of the Si_3N_4 substrate is also evidenced by the reduced thickness of the oxidation layer (2.5 μm) in comparison with dry oxidation at the same temperature and annealing time (15 μm). In this case, the fast degradation of the Si_3N_4 substrate during hot gas corrosion at 1200 °C was likely caused by the high current stream in combination with the low viscosity of the oxidation layer. The remaining of a thin oxide layer could be a consequence of the lower p_{H2O} in the last 75 h of the test, accounting for lower corrosion rates.

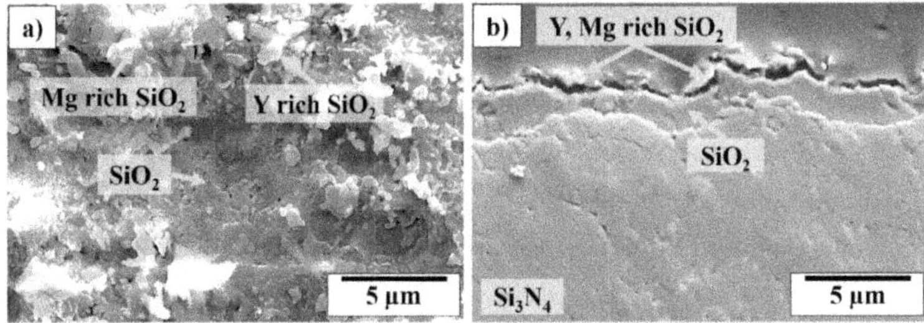

Fig. 4.4.11. SEM micrographs of the (a) surface and (b) cross-section of uncoated Si_3N_4 after hot gas corrosion at 1200 °C for 200 h.

SEM/EDS analysis of the surface of the coated Si_3N_4 after hot gas corrosion shows the elimination of the glassy SiO_2 phase (Fig. 4.4.12 (a)). This explains the higher mass loss in the first 25 h, also increasing the amount of open porosity to 11 vol%, which was estimated by image analysis and is higher than measured for the Yb(2:1) coating (4 vol%) after pyrolysis (Fig. 4.4.1).

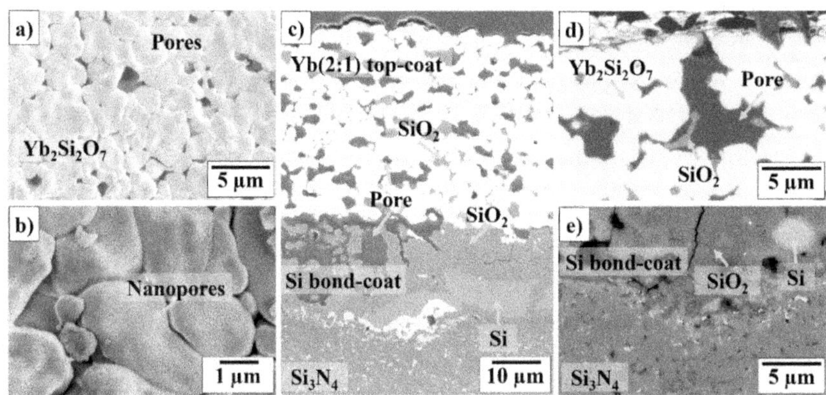

Fig. 4.4.12. SEM micrographs of the (a and b) surface and (c, d and e) cross-section of Si/Yb(2:1) coating system on Si_3N_4 after hot gas corrosion at 1200 °C for 200 h.

The detailed analysis of the $Yb_2Si_2O_7$ grains shows the formation of nanopores (Fig. 4.4.12 (b)), which should be a product of the very low corrosion of $Yb_2Si_2O_7$, thus yielding Yb_2SiO_5, Yb_2O_3 and volatile silanols through the corrosion mechanisms of Eq. 4.4.4 and 4.4.5 [19,25,29,31]. However, neither Yb_2SiO_5 nor Yb_2O_3 were detected by means of XRD (Fig. 4.4.10 (b)) or SEM/EDS.

$Yb_2Si_2O_7 + 2\ H_2O \rightarrow Yb_2SiO_5 + Si(OH)_4$ (Eq. 4.4.4)

$Yb_2SiO_5 + 2\ H_2O \rightarrow Yb_2O_3 + Si(OH)_4$ (Eq. 4.4.5)

The preferred corrosion of the SiO_2 phases also in deeper coating regions is detectable by the analysis of the cross-section, showing the formation of pores up to the interface with the Si_3N_4 substrate (Fig. 4.4.12 (c)). Hence, only residual SiO_2 was still detected by SEM/EDS analysis between the $Yb_2Si_2O_7$ grains as indicated in Fig. 4.4.12 (d). The exposure for 200 h at 1200 °C also led to the formation of some additional cracks in the bond-coat resulting from crystallization effects of SiO_2 (Fig. 4.4.12 (e)).

Whereas the mass loss up to 25 h results from the corrosion of SiO_2 at the surface of the Si/Yb(2:1) coating system, beyond this point and up to 125 h, corrosion is controlled by the inward permeation of moisture into the porous microstructure of the coating and the formation of volatile $Si(OH)_4$ from the SiO_2 phases. Anyhow, this mass loss was overcompensated by the oxidation of residual elemental Si in the bond-coat, which led to a gain in mass up to 125 h. This is supported by the fact that the thickness of the bond-coat has expanded from 10 to 20 μm, following the mechanisms of Eq. 4.4.6 and 4.4.7 (compare Fig. 4.4.12 (e) and Fig. 4.4.2 (a)). The low mass loss within the last 75 h mainly results from the corrosion of SiO_2 from the bond-coat as the silica in the grain boundaries of $Yb_2Si_2O_7$ was already volatilized (see Fig. 4.4.10 (a) and Fig. 4.4.12 (c)).

$Si + O_2 \rightarrow SiO_2$ (Eq. 4.4.6)

$Si + 2\ H_2O \rightarrow SiO_2 + 2\ H_2$ (Eq. 4.4.7)

Finally, the corrosion rate of the coated samples was effectively reduced under these very harsh conditions. A further improvement in the long-term stability of the Si/Yb(2:1) coating system on Si_3N_4 is expected if the formation of the SiO_2 phase between the $Yb_2Si_2O_7$ crystallites can be completely avoided.

4.5 Deposition of thicker Yb(2:1) coatings

Since hot gas corrosion leads to the slow degradation of rare-earth silicate-based environmental barrier coatings, the deposition of thick coatings is desired to extend the life span of coated ceramic parts. The deposition of thicker Yb(2:1) coatings could only be achieved by performing an intermediate pyrolysis step at 750 °C before the deposition of additional layers, according to the procedure described in Section 3.4.1.

At this temperature pyrolysis of Durazane 1800 is almost completed, thus avoiding the excessive volatilization of gaseous species as in the case of thick layers. Moreover, an apparent volume expansion of 2% was determined at 750 °C for the Yb(2:1) monolith (refer to Section 4.2), which was able to comply with the thermal expansion of the Si_3N_4 substrate, yielding stable coatings. Fig. 4.5.1 exhibits the XRD analysis of the resulting Si/Yb(2:1) coating system on Si_3N_4 obtained after 3 cycles in comparison with the coating system from Section 4.4.1, corresponding to 1 cycle.

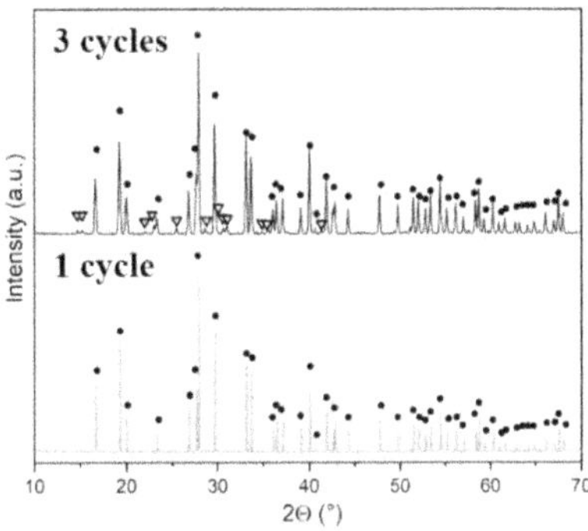

Fig. 4.5.1. XRD analysis of the Si/Yb(2:1) coating system on Si_3N_4 after pyrolysis at 1415 °C for 5 h in air, obtained after 1 and 3 coating cycles (▽ – I2/a Yb_2SiO_5 (00-040-0386); • – β-$Yb_2Si_2O_7$ (00-025-1345)).

Similarly to the coating system obtained after 1 cycle, XRD analysis of the coating obtained after 3 cycles confirmed β-$Yb_2Si_2O_7$ as main crystalline phase. However, the detection of small amounts of I2/a-Yb_2SiO_5 indicates the non-stoichiometric conversion of the Yb_2O_3 filler into the disilicate phase upon reaction with Durazane 1800 and elemental Si powder during pyrolysis. This is in agreement with the phase composition of the corresponding Yb(2:1) monoliths, where traces of the monosilicate phase were still detected embedded in an $Yb_2Si_2O_7$ matrix after pyrolysis at 1415 °C for 5 h in air. More details about the conversion of thicker Yb(2:1) coatings obtained after 3 coating cycles is provided by SEM/EDS analysis, exhibited in Fig. 4.5.2.

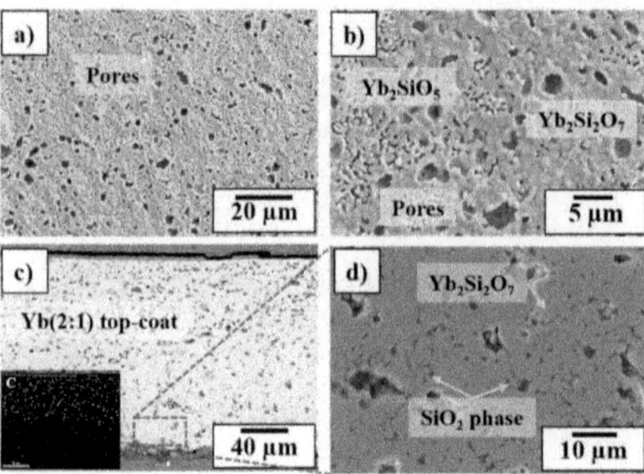

Fig. 4.5.2. SEM micrographs of the (a and b) surface and (c and d) cross-section of the Si/Yb(2:1) coating system on Si_3N_4 after pyrolysis at 1415 °C for 5 h in air, obtained after 3 coating cycles.

By comparing the microstructure of the surface of the coating obtained after 1 cycle (Fig. 4.4.1 (a)) and after 3 cycles (Fig. 4.5.2 (a)), a higher porosity content is evident for the 3 cycle-coating, which was estimated in 11 vol% by image analysis. This result is in agreement with the porosity content of the Yb(2:1) coating after the elimination of the glassy SiO_2 binder phase by hot gas corrosion, which increased from 4 to 11 vol% (Section 4.4.7, Fig. 4.4.12 (a)). EDS analysis of the 3 cycle-coating evidenced the formation of $Yb_2Si_2O_7$ and Yb_2SiO_5 phases (Fig. 4.5.2 (b)), in agreement with XRD measurements (Fig. 4.5.1). In contrast to the monosilicate, the disilicate phase exhibits a dense and compact microstructure. Moreover, no signs of a glassy SiO_2 phase were detected at the surface of the coating by SEM/EDS analysis.

The analysis of its cross-section exhibits a well-adherent layer with a thickness of approximately 150 μm (Fig. 4.5.2 (c)). The EDS mapping of C shows pores, which are concentrated at the surface of the coating system. In contrast, the microstructure closer to the Si_3N_4 substrate is dense (Fig. 4.5.2 (d)). The detailed SEM/EDS analysis of this region confirms the presence of glassy SiO_2 at the grain boundaries of the $Yb_2Si_2O_7$ crystallites with low level impurities of Al, Ca, Na and Mg, corresponding to less than 3 at%, respectively. As previously discussed in Section 4.4.1, the formation and softening of a glassy SiO_2 phase during pyrolysis in air is responsible for the infiltration of the Si/Yb(2:1) coating system and its densification. In the case of thicker coatings, the amount of glassy SiO_2 generated was not sufficient to completely infiltrate its microstructure during pyrolysis at 1415 °C for 5 h in air. Therefore, its accumulation within the grain boundaries of the $Yb_2Si_2O_7$ phase was only detected up to thicknesses of 100 μm from the Si_3N_4 substrate. This also explains the presence of Yb_2SiO_5

exclusively at the surface region of the coating. Despite the residual porosity, the greater thickness and the compact microstructure of the Yb(2:1) coating obtained after 3 cycles should improve the long-term stability of coated ceramic substrates in hot gas environments.

4.6 Si/Yb(2:1) coating system on SiC/SiC

In order to investigate the transferability of the developed Si/Yb(2:1) coating system onto other non-oxide silicon-based ceramics, preliminary coating experiments were carried out with SiC/SiC substrates. Thereafter, the substrates were coated by performing 2 coating cycles, following the procedure for the deposition of thicker Yb(2:1) coatings as described in Section 3.4.1. Fig. 4.6.1 exhibits the XRD analysis of the Si/Yb(2:1) coating system on the SiC/SiC substrate after pyrolysis at 1415 °C for 5 h in air.

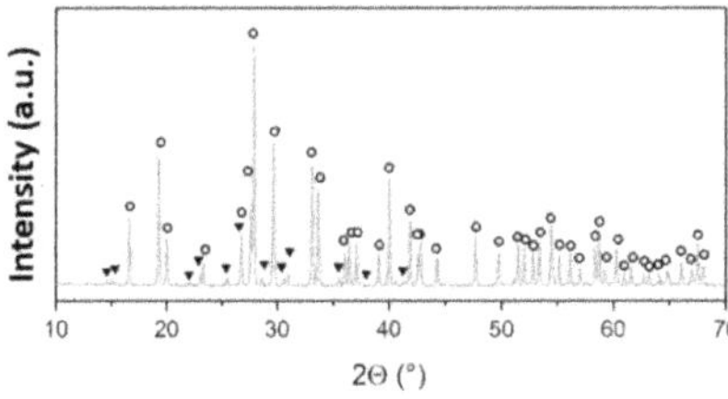

Fig. 4.6.1. XRD analysis of the Si/Yb(2:1) coating system on SiC/SiC after pyrolysis at 1415 °C for 5 h in air, obtained after 2 coating cycles (▼– I2/a Yb_2SiO_5 (00-040-0386); **O** – β-$Yb_2Si_2O_7$ (00-025-1345)).

XRD analysis of the coated SiC/SiC composites detected β-$Yb_2Si_2O_7$ as the main crystalline phase, besides trace amounts of I2/a-Yb_2SiO_5, which is in agreement with the phase composition of the Yb(2:1) monolith after pyrolysis at 1415 °C for 5 h in air (Section 4.2). The optical analysis of the surface of the coated SiC/SiC substrate shows a well-adherent coating with no spallation spots (Fig. 4.6.2 (a)). Nevertheless, the formation of macroscopic longitudinal and vertical cracks is visible at higher magnifications (Fig. 4.6.2 (b)). As determined for the Yb(2:1) monolith, pyrolysis at 1415 °C for 5 h leads to a volume shrinkage of 26%, which in combination with the thermal expansion of the SiC/SiC substrate, resulted in the formation of macroscopic cracks.

The longitudinal and vertical orientation of cracks is possibly a result of the anisotropic thermal expansion of the SiC/SiC substrate in the 45° direction. In the case of Si_3N_4 coated substrates, no cracks were detected in the microstructure of the Yb(2:1) coating by SEM analysis (Section 4.4, Fig. 4.4.1), which was attributed to the infiltration of the coating system by a glassy SiO_2 phase during pyrolysis, relaxing stresses at high temperatures. In agreement with XRD measurements, EDS analysis of the surface of the coating system on SiC/SiC detected $Yb_2Si_2O_7$ and traces of Yb_2SiO_5 (Fig. 4.6.2 (c)). Unlike described for Si_3N_4 substrates, the formation of a glassy SiO_2 phase and the infiltration of the coating system cannot be expected for SiC and SiC/SiC substrates due to the very low or inexistent content of sintering additives. Therefore, an open porosity content of 15 vol% was determined by image analysis, which is higher than the Si/Yb(2:1) coating system on Si_3N_4 (4 vol%, refer to Section 4.4.1).

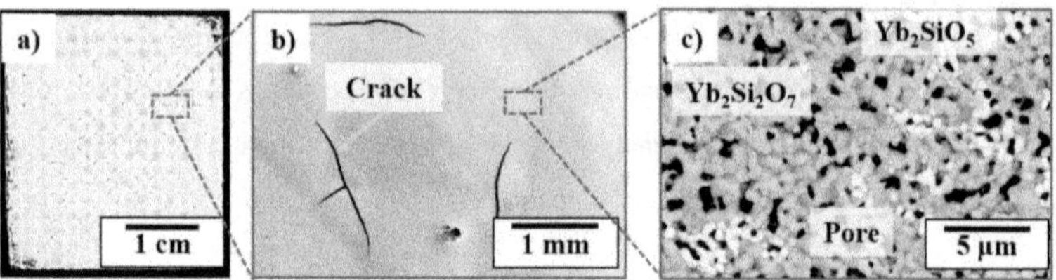

Fig. 4.6.2. Optical and corresponding SEM micrographs of the surface of the Si/Yb(2:1) coating system on SiC/SiC after pyrolysis at 1415 °C for 5 h in air, obtained after 2 coating cycles.

The SEM/EDS analysis of the cross-section of the Si/Yb(2:1) coating system on SiC/SiC evidences a well-adhered and 120 μm thick layer (Fig. 4.6.3). The detailed analysis of regions 1, 2 and 3 is exhibited in Fig. 4.6.4.

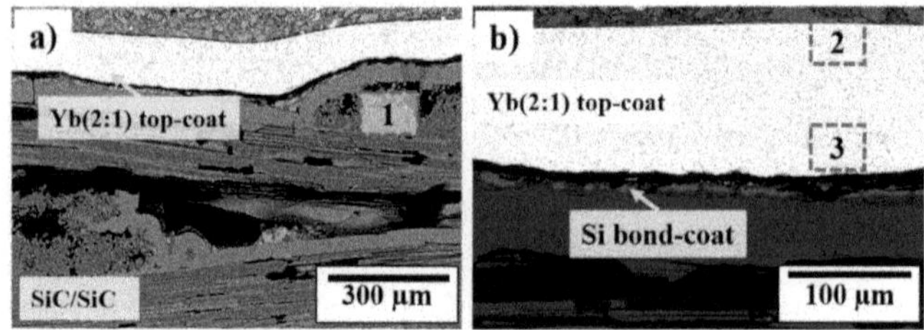

Fig. 4.6.3. SEM micrographs of the cross-section of the Si/Yb(2:1) coating system on SiC/SiC after pyrolysis at 1415 °C for 5 h in air, obtained after 2 coating cycles.

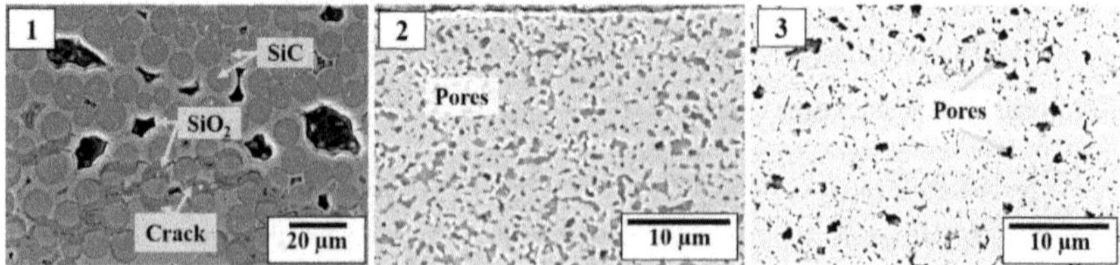

Fig. 4.6.4. Detailed SEM/EDS analysis of regions 1, 2 and 3 from Fig. 4.6.3.

EDS spot analysis of region 1 evidenced the oxidation of the SiC fibers during pyrolysis at 1415 °C for 5 h in air. It is known that the densification of the Si/Yb(2:1) coating system only occurs at 1415 °C as discussed in Section 4.2, thus enabling the diffusion of oxygen and the oxidation of the SiC fibers, which was favored by the residual porosity content of the SiC/SiC substrate, estimated in 10 vol% (Section 3.1.1.2). The oxidation of SiC into SiO_2 is accompanied by a molar volume expansion of 107% [119], forming the cracks seen in Fig. 4.6.4, region 1.

The detailed analysis of regions 2 and 3 evidences two different microstructures. In contrast to region 2, the microstructure closer to the SiC/SiC substrate is dense and compact (region 3). In comparison with free-standing samples as the Yb(2:1) pressed monoliths, the adhesion of the coating system to the substrate hinders the densification of the $Yb_2Si_2O_7$ phase, whereas only shrinkage across the thickness may occur freely. In this case, the further densification of the Si/Yb(2:1) coating system could be expected by increasing the annealing time during pyrolysis. Nonetheless, the good transferability of the developed Si/Yb(2:1) coating system to other non-oxide Si-based ceramics has been demonstrated. As will be mentioned in Section 6, the additional use of corrosion resistant ceramic glass fillers based on the system $Yb_2O_3/Al_2O_3/SiO_2$ can help densify the microstructure of the Yb(2:1) coating during pyrolysis, enabling the deposition of fully dense coating systems to protect ceramic components during long-term operation at temperatures above 1400 °C.

5 Summary

The next generation of gas turbines is expected to operate with inlet temperatures that exceed 1400 °C, thus achieving efficiencies of above 63% and reduced greenhouse gas emissions. Long-term service at these high temperatures is only possible by the substitution of the currently used Ni-based superalloy components by Si_3N_4, SiC and SiC/SiC ceramics and composites. However, water vapor is one of the main products of combustion, which rapidly reacts with these ceramics at temperatures above 1200 °C, forming volatile silanols and leading to surface recession. A reliable protection can be achieved by the deposition of environmental barrier coatings (EBCs) based on ytterbium disilicate ($Yb_2Si_2O_7$), which stands out due to a high temperature endurance of at least 1400 °C, very low corrosion rate, low CTE (3.91 10^{-6} K^{-1}) and no polymorphic transformation.

In general, EBCs for non-oxide silicon-based ceramics are deposited by air plasma spray, chemical (CVD) or physical vapor deposition (PVD), whose main disadvantages rely on the costly procedure besides the difficult coating of complex geometries. As an alternative to these methods, the polymer derived ceramic (PDC) route offers a straightforward approach for the deposition of ceramic coatings after pyrolysis, by typical polymer processing techniques as spraying or dip-coating. Among the available precursors for coating application, silazanes are commercially available, possess a high ceramic yield and ensure a strong adhesion to ceramic substrates as well as a high stability in oxidative and corrosive media. However, the protective effect of silazane-based coatings in oxidative environments is usually based on the formation of a passivating SiO_2-scale, which is rapidly degraded in combustion environments. Thus, one promising approach to increase their stability is the conversion of the forming SiO_2 into ytterbium silicates by the use of Yb_2O_3 as an active filler.

The development of the $Yb_2Si_2O_7$ coatings started by understanding the role of silazanes in the conversion of Yb_2O_3 into ytterbium silicates upon pyrolysis. In preliminary studies, monoliths were processed from the oligosilazane Durazane 1800 and Yb_2O_3 in different molar ratios to yield ytterbium mono (Yb_2SiO_5) or disilicates after pyrolysis in air. The processing and pyrolysis of additional monoliths in argon atmosphere led to the formation of $Yb_4Si_2N_2O_7$ as main crystalline phase, which resulted in the collapse of dense samples upon oxidation at high temperatures due to the considerable volume expansion associated with the conversion of this phase into ytterbium mono- and disilicates. Thus, pyrolysis in air was proven to be imperative to obtain stable systems. Due to the shrinkage of the silazane precursor during pyrolysis, first crack-free but porous coatings could only be obtained by limiting its fraction.

Consequently, the amount of silicon available in the coating was not sufficient to quantitatively convert the Yb_2O_3 filler into the respective silicate phases. Therefore, elemental Si powder was added as an additional active filler to achieve the stoichiometric conversion of Yb_2O_3 into $Yb_2Si_2O_7$.

To acquire information on the formation and the sintering behavior of the $Yb_2Si_2O_7$ phase, any influence of the substrate was excluded by investigating powder and monolithic samples processed from the developed Yb(2:1) slurry. TG analysis of the Yb(2:1) composite powder determined a mass gain of 105 wt% after pyrolysis at 1415 °C for 5 h in air, whereas a theoretical yield of 97 wt% was calculated based on the mass change of the individual Yb(2:1) components. This provided evidence for a higher reactivity of Durazane 1800 and elemental Si powder with oxygen from the atmosphere in the presence of Yb_2O_3. The investigation of Yb(2:1) monoliths allowed the elucidation of the phase composition and microstructure of the resulting ytterbium silicates. As confirmed by XRD and SEM/EDS analyses, the ytterbium silicate phases started forming at above 1000 °C in air. At 1300 °C, β-$Yb_2Si_2O_7$ was already the main crystalline phase and only remaining traces of Yb_2SiO_5 were detected. The pressureless sintering of the generated β-$Yb_2Si_2O_7$ phase at 1415 °C resulted in a volumetric shrinkage of 26%, regarding the corresponding Yb(2:1) monolith before pyrolysis. The remarkable densification of the pressed monoliths was confirmed by the low open porosity content of 4.5 vol% and a density of 5.95 g cm^{-3} after pyrolysis at 1415 °C for 5 h, corresponding to approximately 97% of the bulk density of β-$Yb_2Si_2O_7$ (6.15 g cm^{-3}). Annealing at higher temperatures and holding times did not result in any significant microstructural changes. Therefore, the pyrolysis temperature of 1415 °C and 5 h in air was selected to generate dense and crystalline β-$Yb_2Si_2O_7$ coatings.

Preliminary research with the developed Yb(2:1) coatings on Si_3N_4 revealed the limited oxidation stability of the selected substrates at above 1400 °C in air. The high temperatures necessary to yield dense silicate phases led to the enhanced oxidation of the Si_3N_4 substrate and the formation of blisters, which compromised the mechanical stability of the deposited coatings. In this context, the oxidation behavior of uncoated Si_3N_4 substrates was investigated up to 1415 °C in air, which allowed the detailed description of the conversion behavior, phase composition and microstructure of the Yb(2:1) coatings. As confirmed by XRD and SEM/EDS analyses, the formation of a passivating SiO_2 layer at the surface of the Si_3N_4 substrate during oxidation created a driving force for the diffusion of sintering additives and cation impurities from the bulk, responsible for decreasing its viscosity and enhancing oxidation. This resulted in the accumulation of Ca, Al, Y, Mg and Na metals in the oxide layer, besides the precipitation of $Y_2Si_2O_7$ and $MgSiO_3$ silicates phases and the formation of glassy SiO_2 phases.

The enhanced oxidation of the Si_3N_4 substrate during pyrolysis and the formation of blisters were effectively avoided by the development and deposition of a Si bond-coat based on Durazane 2250 and elemental Si powder. After the application of the bond-coat, the deposition of the β-$Yb_2Si_2O_7$ coating occurred by spraying the Yb(2:1) slurry as a top-coat, followed by pyrolysis at 1415 °C for 5 h in air. This resulted in a well-adherent, dense and 68 μm thick coating system, named as Si/Yb(2:1). XRD analysis of the Si/Yb(2:1) coating system on Si_3N_4 evidenced β-$Yb_2Si_2O_7$ as only crystalline phase. Nevertheless, the accumulation of a glassy SiO_2 phase at the grain boundaries of the β-$Yb_2Si_2O_7$ crystallites was confirmed by SEM/EDS analysis, which can be explained by the formation of a passivating SiO_2 layer at the interface between the bond-coat and the Si_3N_4 substrate. The diffusion of sintering additives and metal impurities from the bulk of the substrate into it and the consequent reduction of viscosity favored the infiltration of the coating system with a glassy SiO_2 phase. Besides contributing to the conversion of the Yb_2O_3 filler into β-$Yb_2Si_2O_7$, it resulted in a dense microstructure with an open porosity content of 4 vol% as determined by image analysis.

The good mechanical properties of the resulting coating system also benefited from the strong interaction with the Si_3N_4 substrate during pyrolysis in air and its infiltration with a glassy SiO_2 phase. Pull-off tests measured an outstanding adhesion of 36.9 ± 6.2 MPa. SEM/EDS of the tested area revealed that the bond-coat failed by cohesion, possibly caused by the presence of small pores and cracks after pyrolysis. In addition, microhardness measurements at the surface of the coating system determined a hardness of 713 ± 169 (HV 0.1) (6.9 ± 1.6 GPa), comparable to the intrinsic hardness of $Yb_2Si_2O_7$ (7.3 ± 0.2 GPa), resulting from the high density of the Yb(2:1) coatings after pyrolysis at 1415 °C for 5 h in air. As a complement to pull-off and hardness measurements, scratch tests determined a critical load of 28 N necessary to damage the surface of the coating system, leading to spallation and the formation of wedges, which is typical of well-adherent hard coatings on hard substrates, where plastic deformation is minimal and fracture dominates the scratch response. In comparison with the uncoated Si_3N_4 substrate, a flexural strength of 547 ± 20 MPa was measured for the Si/Yb(2:1) coated Si_3N_4, which accounts for a debit in strength of 32% at room temperature. The debit in strength was caused by residual thermal stresses due to the CTE mismatch between coating system and substrate, which was intensified by the strong bonding of the coating system and the interaction of its flaws with pre-existing ones in the Si_3N_4 ceramic substrates.

Thermal cycling carried out between 1200 and 20 °C by quenching in a water bath and hot gas corrosion test at 1200 °C for 200 h (p = 1 atm, p_{H2O} = 0.15 atm and v = 100 m s^{-1}) evaluated the performance of the developed coating system, simulating the very harsh environments in gas turbines. The good CTE compatibility between the coating system and the Si_3N_4 substrate resulted in an increased thermal shock resistance. Thus, it remained fully adhered up to 6 cycles, whereas only a few spallation spots were seen at the edge of the coated substrate after 15 cycles. The good performance during thermal shock tests was also attributed to the residual porosity within the microstructure of the coating and to the softening of the glassy SiO_2 phase at high temperatures. The residual porosity decreases the Young's modulus of the coating, thus increasing its thermal shock tolerance, whereas the softening of the glassy SiO_2 phase during annealing at 1200 °C likely led to the relaxation of stresses generated upon thermal cycling. In comparison with uncoated Si_3N_4 substrates, the Si/Yb(2:1) coating system effectively reduced the mass loss caused by hot gas corrosion in 98% after 200 h. Thereby, the coated samples appeared intact and the coatings remained well-adhered to the Si_3N_4 substrate. Nevertheless, SEM/EDS analysis evidenced the preferable corrosion of the glassy SiO_2 phase within the grain boundaries of the β-$Yb_2Si_2O_7$ phase. Therefore, hindering its formation would impact positively in the long-term stability of the coated Si_3N_4 substrates. The high density of the generated $Yb_2Si_2O_7$ phase, its minor degradation during hot gas corrosion and the excellent performance of the developed EBC system during thermal cycling confirm its potential for the protection of other non-oxide silicon-based ceramics in extreme combustion environments.

At last, work was carried out on the deposition of thicker Yb(2:1) coatings and on the coating of SiC/SiC substrates. Thereby, a coating thickness of 150 μm was achieved, which should contribute to increase the life-span of coated ceramic parts in combustion environments. In addition, the good transferability of the developed Si/Yb(2:1) coating system onto other non-oxide Si-based ceramics has been demonstrated.

6 Outlook

In this work, the potential of the developed Si/Yb(2:1) coating system for protection of non-oxide silicon-based ceramics was confirmed after thermal cycling and hot gas corrosion tests. Nevertheless, the generation of a dense β-$Yb_2Si_2O_7$ coating during pyrolysis at 1415 °C for 5 h in air led to the oxidation of the interface between the bond-coat and the selected Si_3N_4 substrate. This resulted in the infiltration of the coating system with a glassy SiO_2 phase and its accumulation at the grain boundaries of the β-$Yb_2Si_2O_7$ crystallites. During hot gas corrosion, the glassy SiO_2 phase was preferably corroded, leading to the slow degradation of the coating system and compromising its long-term stability. In order to enable the suitable protection of the Si_3N_4 substrate in combustion environments for considerable amount of times ($t > 10{,}000$ h) as required for gas turbines, the formation and accumulation of glassy SiO_2 at the grain boundaries of the β-$Yb_2Si_2O_7$ phase should be hindered.

The disparity in concentration of sintering additives and impurities within the grain boundaries of the Si_3N_4 substrate and the formed passivating SiO_2 layer at the interface with the bond-coat is the driving force for their diffusion, which results in the development of glassy SiO_2 [78]. Thus, one way to suppress its formation could be the application of a bond-coat with a suitable concentration of the respective oxides from the metals detected in the glassy SiO_2 phase within the Yb(2:1) coating after pyrolysis at 1415 °C for 5 h in air (e.g. Ca, Mg, Al and Na). Alternatively, the deposition of an intermediary layer rich in Yb_2O_3 between the bond-coat and Yb(2:1) top-coat could avoid the accumulation of glassy SiO_2 at the grain boundaries of the β-$Yb_2Si_2O_7$ phase. Thereby, the infiltration of the coating system with glassy SiO_2 during pyrolysis should enable the conversion of Yb_2O_3 within the intermediary layer into $Yb_2Si_2O_7$. Ueno *et al.* [92] adopted the same strategy during the deposition of $Lu_2Si_2O_7$ coatings onto Si_3N_4 substrates. For this purpose, the authors applied a bond-coat rich in Lu_2O_3, enabling its conversion into $Lu_2Si_2O_7$ during pyrolysis, which has a matching CTE to Si_3N_4.

A second strategy consists in converting the glassy grain boundary SiO_2 phase into corrosion resistant rare-earth silicates. This could be achieved by the addition of rare-earth metal or oxide powders with a nanoscale particle size as additives to the developed Yb(2:1) slurry. A similar approach is already adopted for the processing of Si_3N_4 ceramics. During the sintering of Si_3N_4, a glassy grain boundary SiO_2 phase is formed, whereas ions are able to diffuse from the sintering additives with a nanoscale particle size (< 100 nm) as Yb_2O_3, Y_2O_3 and Sc_2O_3 and convert the amorphous SiO_2 grain boundary phase into the respective silicates [19,77,156]. Another suitable approach could be the use of rare-earth modified silazane precursors, which should substitute the currently used silazane Durazane 1800 as a binder. The modification of silazanes with transition metals and rare-earths was already researched in detail [157–164]. Thereby, different metals (e.g. Cu, Pd, Hf and Yb) were chemically introduced into the backbone of silazanes by the reaction with NH and SiH functional groups. After cross-linking, pyrolysis of the metal modified silazanes at temperatures above 900 °C led to the formation of the respective oxides, nitrides, silicides or silicate phases. The advantage of this approach is the targeted and homogeneous distribution of rare-earths in an atomic level at the grain boundaries of the $Yb_2Si_2O_7$ phase, thus enabling their conversion into the respective silicate phases.

Besides the formation of the glassy SiO_2 phase, the Yb(2:1) top-coat presented a residual porosity of 11 vol% after hot gas corrosion, which may facilitate the permeation of moisture and the corrosion of the coated system. In this case, the full densification of the coating system should also contribute to its long-term stability in hot gas environments. Thus, the use of corrosion resistant meltable ceramic fillers should help densify its microstructure and increase the stability of the coating system in combustion environments. In this regard, Vogt and Nöth [165] developed a slurry-based yttrium aluminosilicate (YAS) coating for monolithic SiC. Thereafter, coated SiC substrates were tested in hot gas environments at 1200 °C for 200 h ($p = 1$ atm, $p_{H2O} = 0.15$ atm, $v = 100$ m s^{-1}), whereby the coating successfully reduced mass loss by corrosion in comparison with uncoated SiC substrates. The same sort of fillers could be employed in the Si/Yb(2:1) coating system. In comparison with YAS [166], ytterbium aluminosilicates (YbAS) possess a higher eutectic temperature. In the YbAS system, the lowest melting point is reported at 1500 °C, which is dependant on the ratio among the Yb_2O_3, Al_2O_3 and SiO_2 powder components used for its synthesis [167]. The melting of this filler should densify the microstructure of the Yb(2:1) coating during pyrolysis, thus enabling the deposition of fully dense coating systems for protecting ceramic components during operation at temperatures above 1400 °C.

A debit in the flexural strength of 32% was caused by the deposition and pyrolysis of the Si/Yb(2:1) coating system at room temperature, regarding uncoated Si_3N_4 substrates, which can be overcome by changing the coating architecture. In this regard, Bhatia *et al.* [26,36] suggested the application of a low modulus interlayer between the Si bond-coat and the Si_3N_4 substrate. By depositing a porous oxide interlayer, the coated Si_3N_4 substrates retained the flexural strength at room temperature, which hindered the propagation of cracks from the coating into the substrate during 4-point-bending tests. Nevertheless, the introduction of a porous interlayer in addition to the Si/Yb(2:1) coating system is expected to compromise the coating adhesion and decrease its scratch tolerance.

Finally, in comparison with ytterbium disilicate, ytterbium monosilicate possesses an increased stability in combustion environments [19]. Therefore, one potential strategy to achieve an increased protection against hot gas corrosion is the deposition of a tri-layer coating system consisting of a Si-based bond-coat, $Yb_2Si_2O_7$ intermediary layer and Yb_2SiO_5 top-coat. In this regard, the mol ratio between Si and Yb_2O_3 available in the Yb(2:1) slurry can be adjusted to yield the monosilicate phase instead of the disilicate. Thereby, the volume fraction of Durazane 1800 must be optimized to avoid its excessive shrinkage during pyrolysis and enable the deposition of well-adherent and stable coatings. Nevertheless, due to the different CTEs between the monosilicate (Yb_2SiO_5, CTE = 6.65 10^{-6} K^{-1} [19]) and the other coating layers ($Yb_2Si_2O_7$, CTE = 3.91 10^{-6} K^{-1} [19]; Si, CTE = 4.1 10^{-6} K^{-1} [27]; Si_3N_4, CTE = 3.60 10^{-6} K^{-1} [19]), a reduced thermonechanical stability and the formation of a cracked microstructure can be expected during thermal shock tests, similarly as reported for the tri-layer system Si bond-coat, mullite intermediary layer (CTE = 5.71 10^{-6} K^{-1} [19]) and Yb_2SiO_5 top-coat [10,22,91].

7 Zusammenfassung

Bei der nächsten Generation von Gasturbinen steht ein langfristiger Betrieb bei Temperaturen oberhalb von 1400 °C zur Steigerung des Wirkungsgrades auf über 63 % und zur Reduzierung der Treibhausgasemissionen im Vordergrund. Dies ist allerdings nur durch die Ersetzung der aktuell in Gasturbinen eingesetzten Nickelbasissuperlegierungen durch temperaturbeständigere Werkstoffe möglich. Zu solchen Materialien zählen vor allem die keramischen Werkstoffe Si_3N_4, SiC und SiC/SiC. An Luftsauerstoff bildet sich bei diesen nichtoxidischen Keramiken eine passivierende SiO_2-Schicht aus. Ab 1200 °C tritt eine Reaktion des in den Verbrennungsgasen enthaltenen Wasserdampfes mit der Oxidschicht ein, die zur Korrosion und folglich zu einer schnellen Degradation der Oberfläche der Keramikbauteile führt.

Deshalb sind sogenannte „Environmental Barrier Coatings (EBCs)" notwendig, um einen geeigneten Schutz dieser Keramiken bei entsprechenden Bedingungen zu gewährleisten. Als ein vielversprechendes EBC-Material zeichnet sich Ytterbiumdisilikat ($Yb_2Si_2O_7$) aus. Grund hierfür sind die hohe Temperaturbeständigkeit bis mindestens 1400 °C, eine geringe Korrosionsrate, ein zu den nichtoxidischen Keramiken passender thermischer Ausdehnungskoeffizient (3,91 10^{-6} K^{-1}) und keine polymorphe Umwandlung. EBCs für Si_3N_4, SiC und SiC/SiC werden meistens mittels kostenintensiven und apparativ aufwendigen Verfahren, wie z.B. Plasmaspritzen, chemische (CVD) oder physikalische (PVD) Gasphasenabscheidung appliziert. Im Gegensatz dazu stellen schlickerbasierte Methoden, wie z.B. die PDC-Route (polymer derived ceramics), eine kostengünstige und alternative Methode zur Herstellung keramischer Schichten dar. Zudem bietet die Precursortechnik die Möglichkeit, Schichten durch einfache Verfahren, wie z.B. Tauchen oder Sprühen, auch auf komplexe Geometrien zu applizieren. Aufgrund der hohen keramischen Ausbeute und der guten Haftung auf keramischen Substraten zeichnen sich insbesondere Silazane als Beschichtungsmaterial aus. Allerdings bilden auch die Silazanbeschichtungen in oxidativer Atmosphäre eine passivierende SiO_2-Schicht aus, die in Verbindung mit Wasserdampf degradiert. Zur Erhöhung der Stabilität der Silazanbeschichtungen in Verbrennungsatmosphären wurde Ytterbiumoxid (Yb_2O_3) als aktiver Füllstoff verwendet, der durch eine Reaktion mit SiO_2 heißgaskorrosionsbeständige Silikate ausbildet.

Die grundlegende Entwicklung von $Yb_2Si_2O_7$-Schichten auf Basis der PDC-Technologie erfolgte durch die Untersuchung der Wechselwirkung zwischen Silazanen und Yb_2O_3. Deshalb wurden in Vorarbeiten Monolithen aus dem Oligosilazan Durazane 1800 und Yb_2O_3 hergestellt, wobei das Molverhältnis zwischen beiden Edukten variiert wurde, um gezielt Ytterbiummono- (Yb_2SiO_5) oder -disilikat bei der reaktiven Pyrolyse an Luft zu bilden. Im Gegensatz dazu führte die Pyrolyse dieser Monolithen in einer Argonatmosphäre zur Bildung von $Yb_4Si_2N_2O_7$. Diese Phase führte bei einer anschließenden Oxidation zur Bildung von Ytterbiumsilikaten und zur Zerstörung des gepressten Monolithen, was auf die damit verbundene Volumenzunahme zurückzuführen ist. Diese Untersuchungen verdeutlichten, dass zur Erzeugung stabiler Silikatphasen eine Pyrolyse an Luft notwendig ist. Aufgrund der hohen Volumenschrumpfung des Silazans während der Pyrolyse und der damit verbundenen Rissbildung, war der Anteil an Durazane 1800 im Beschichtungssystem limitiert, das allerdings in einer unvollständigen Umwandlung von Yb_2O_3 in $Yb_2Si_2O_7$ resultierte. Deshalb wurde elementares Si-Pulver als zusätzlicher Füllstoff und Siliziumquelle zu den Suspensionen hinzugefügt. Dadurch konnte eine vollständige Bildung von $Yb_2Si_2O_7$ gewährleistet werden und gleichzeitig die Schrumpfung während der Pyrolyse minimiert werden.

Die grundlegenden Untersuchungen zur Bildung der $Yb_2Si_2O_7$-Phase und zum Sinterverhalten erfolgten ohne Beteiligung eines Substrats. Hierzu wurde ausgehend von der entwickelten Yb(2:1)-Suspension ein Pulver hergestellt, aus dem u.a. wiederum monolithische Probekörper gepresst wurden. Bei der TG-Analyse des Yb(2:1)-Pulvers trat eine Massenzunahme um 5 Gew.% nach einer Pyrolyse bei 1415 °C für 5 h an Luft auf, wobei auf Basis der Massenänderung der jeweiligen Yb(2:1)-Komponenten eine theoretische Ausbeute von 97 Gew.% erreicht wurde. Dies deutete auf eine hohe Reaktivität zwischen Durazane 1800 bzw. Si mit Yb_2O_3 hin, die zur Bildung der Silikatphasen an Luft führt. XRD- und REM/EDX-Untersuchungen an den Yb(2:1)-Monolithen bestätigten die Bildung von Ytterbiumsilikatphasen bei einer thermischen Auslagerung ab 1000 °C an Luft. β-$Yb_2Si_2O_7$ bildete sich als Hauptkristallphase bereits nach einer Pyrolyse bei 1300 °C aus. Infolgedessen wurden nur noch geringe Spuren von Yb_2SiO_5 nach einer Pyrolyse bei 1415 °C nachgewiesen. Darüber hinaus trat bei dieser Temperatur das Versintern der β-$Yb_2Si_2O_7$-Phase auf, das mit einer Volumenschrumpfung von 26%, bezogen auf das Volumen des entsprechenden Yb(2:1)-Monoliths vor der Pyrolyse, verbunden ist. Der Monolith wies nach einer Pyrolyse bei 1415 °C für 5 h eine offene Porosität von 4,5 Vol.% und eine Dichte von 5,95 g cm^{-3} auf, das zu 97% der Raumdichte von β-$Yb_2Si_2O_7$ entspricht (6,15 g cm^{-3}). Eine höhere Pyrolysetemperatur oder längere Haltezeiten führte zu keiner weiteren Änderung der Mikrostruktur oder Phasenzusammensetzung der Yb(2:1)-Monolithen. Aufgrund dessen eigneten sich für die quantitative Umwandlung von Durazane 1800, Yb_2O_3 und elementarem Si in dichtes und kristallines β-$Yb_2Si_2O_7$ eine Pyrolyse an Luft bei einer Temperatur von 1415 °C und einer Haltezeit von 5 h.

Voruntersuchungen mit dem entwickelten Yb(2:1)-System zeigten nur eine geringe Oxidationsschutzwirkung der Si_3N_4-Substrate bei 1400 °C an Luft, was sich in der Bildung von Blasen in der Beschichtung während der Pyrolyse äußerte. Um das Oxidationsverhalten des verwendeten Si_3N_4-Substrats näher zu charakterisieren, wurden an unbeschichteten Substraten Oxidationstests bis 1415 °C an Luft durchgeführt. Die Bildung einer passivierenden SiO_2-Schicht während der Oxidation resultierte in der Diffusion der im Si_3N_4 enthaltenden Sinteradditiven und Verunreinigungen an die Oberfläche und folglich zu einer Anreicherung der Substratoberfläche mit Ca, Al, Y, Mg und Na. Infolgedessen wurde mittels REM/EDX- und XRD-Analysen neben der niedrigschmelzenden amorphen SiO_2-Glasphase, auch die Kristallisation von SiO_2 (α-Cristobalit), $Y_2Si_2O_7$- und $MgSiO_3$-Silikatphasen nachgewiesen.

Die starke Oxidation der Si_3N_4-Substrate während der Pyrolyse der Yb(2:1)-Schicht ließ sich durch die Entwicklung und Applizierung eines Bond-Coats auf Basis von Durazane 2250 und elementarem Si-Pulver vermeiden. Die anschließende Pyrolyse der Yb(2:1)-Schicht bei 1415 °C für 5 h an Luft resultierte in einem gut haftenden, dichten (4 Vol.% Restporosität) und 68 μm dicken Schichtsystem (sog. „Si/Yb(2:1)“), bestehend aus Bond- und Top-Coat, wobei β-$Yb_2Si_2O_7$ als einzige kristalline Phase mittels XRD nachgewiesen werden konnte. Dennoch wurde eine Anreicherung einer SiO_2-Glasphase an den Korngrenzen der β-$Yb_2Si_2O_7$-Phasen mittels REM/EDX nachgewiesen. Die Bildung der SiO_2-Glasphase ist auf die Oxidation der Substratoberfläche unter dem Bond-Coat zurückzuführen. Die anschließende Diffusion von im Si_3N_4-Substrat enthaltenden Sinteradditiven und Verunreinigungen in die gebildete passivierende SiO_2-Schicht führte wiederum zur Erniedrigung ihres Schmelzpunkts der SiO_2-Phase. Demzufolge erfolgte die Infiltration des Schichtsystems mit dieser Phase, wobei der Überschuss an SiO_2 zur Verdichtung der Yb(2:1)-Schicht und zudem zur Umwandlung von Yb_2O_3 in $Yb_2Si_2O_7$ beitrug.

Zusätzlich hatte die Infiltration der Schicht mit der SiO_2-Glasphase einen positiven Einfluss auf die mechanischen Eigenschaften des Schichtsystems, wodurch u.a. die Haftung der Beschichtung am Substrat verbessert wurde. Die mittels Pull-Off-Test bestimmte Haftfestigkeit betrug 36,9 ± 6,2 MPa. REM/EDX-Analysen vom getesteten Bereich deuteten auf ein Versagen im Bond-Coat hin, was auf die Bildung von Mikrorissen und Poren im Bond-Coat während der Pyrolyse zurückzuführen ist. Aufgrund der hohen Schichtdichte wurde eine hohe Härte von 713 ± 169 (HV 0,1) (6,9 ± 1,6 GPa) ermittelt, die mit der Härte von reinem $Yb_2Si_2O_7$ (7,3 ± 0,2

GPa) vergleichbar ist. Zusätzlich zur Haftfestigkeit und Härte wurde die Ritzbeständigkeit der pyrolysierten Beschichtung untersucht. Die Schicht hielt einer Belastung von 28 N stand. Höhere Lasten führten zu Abplatzungen und keilförmigen Verschleißspuren, das das typische Schadensmuster von harten Schichten auf harten Substraten darstellt. Aufgrund der Mikrorisse und Poren im Schichtsystem und der starken Haftung zum Substrat nahm die Biegefestigkeit der beschichteten Si_3N_4-Proben im Vergleich zu den unbeschichteten Si_3N_4-Substraten (799 ± 47 MPa) bei Raumtemperatur um 32 % ab.

Mittels Thermoschockversuchen durch Abschrecken in Wasser von 1200 auf 20 °C und durch Heißgaskorrosionstests bei 1200 °C für 200 h (p = 1 atm, p_{H2O} = 0,15 atm und v = 100 m s^{-1}) wurde das Potenzial des entwickelten Schichtsystems unter praxisnahen Bedingungen, wie sie in Gasturbinen vorherrschen, getestet. Aufgrund der guten Kompatibilität im thermischen Ausdehnungskoeffizienten zwischen Schichtsystem und Substrat, verblieb die Beschichtung bis zum sechsten thermischen Zyklus auf dem Substrat vollständig haften, wobei nach 15 Zyklen nur wenige kleinflächige Abplatzungen an den Rändern des Substrats zu beobachten waren. Zum einen ist dieses gute Thermoschockverhalten der Schicht auf die Restporosität zurückzuführen, die den E-Modul der Schicht lokal herabsetzt und somit die Thermoschockbeständigkeit erhöht. Zum anderen führte die Erweichung der SiO_2-Glasphase bei 1200 °C zur Verringerung des Spannungszustandes in der Schicht. Beim Heißgaskorrosionstest zeigte sich, dass durch das Schichtsystem der Massenverlust um 98% im Vergleich zu den unbeschichteten Si_3N_4-Substraten reduziert werden konnte. Mittels REM/EDX-Untersuchungen konnte nachgewiesen werden, dass die Korrosion bevorzugt in der SiO_2-Glasphase an den Korngrenzen der β-$Yb_2Si_2O_7$-Phase stattfindet. Dies kann wiederum die Langzeitstabilität des Schichtsystems negativ beeinträchtigen. Aus diesem Grund ist die Bildung dieser Glasphase zu unterbinden, um eine höhere Stabilität des beschichteten Si_3N_4-Substrats bei Heißgaskorrosionsbedingungen zu erzielen. Zusammenfassend lässt sich sagen, dass sich das entwickelte Schichtsystem aufgrund der hohen Dichte, der hohen Stabilität der erzeugten $Yb_2Si_2O_7$-Phase in Heißgaskorrosionsumgebungen und des hervorragenden Thermoschockverhaltens, auch einen vielversprechenden Ansatz zum Schutz vor Oxidation und Korrosion anderer nichtoxidischer Si-basierten Keramiken liefert.

Schließlich wurden grundlegende Arbeiten über die Applizierung dickerer Yb(2:1)-Schichten durchgeführt, wobei eine Schichtdicke von 150 μm erreicht wurde, was die Langzeitstabilität des entwickelten Schichtsystems in Heißgaskorrosionsumgebungen verbessern soll. Darüber hinaus wurde die gute Übertragbarkeit des Beschichtungssystems auf SiC/SiC demonstriert.

8 References

[1] CO_2 Emissions from Fuel Combustion, Paris - France, 2020. https://www.iea.org/reports/co2-emissions-from-fuel-combustion-overview, (accessed Feburary 11, 2021).

[2] Energy Technology Perspectives 2012, Paris - France, 2012. https://www.iea.org/reports/energy-technology-perspectives-2012, (accessed February 11, 2021).

[3] Global Energy & CO_2 Status Report 2019, Paris - France, 2019. https://www.iea.org/reports/global-energy-co2-status-report-2019, (accessed Feburary 11, 2021).

[4] B. McLellan, N. Florin, D. Giurco, Y. Kishita, K. Itaoka, T. Tezuka, Decentralised energy futures: The changing emissions reduction landscape, in: Procedia CIRP, Elsevier B.V., 2015: pp. 138–143. doi:10.1016/j.procir.2015.02.052.

[5] H. Kobayashi, A. Hayakawa, K.D.K.A. Somarathne, E.C. Okafor, Science and technology of ammonia combustion, Proc. Combust. Inst. 37 (2019) 109–133. doi:10.1016/j.proci.2018.09.029.

[6] M.P. Boyce, Gas Turbine Engineering Handbook, Elsevier Inc., 2012. doi:10.1016/C2009-0-64242-2.

[7] T.M. Pollock, S. Tin, Nickel-Based Superalloys for Advanced Turbine Engines: Chemistry, Microstructure and Properties, J. Propuls. Power. 22 (2006) 361–374. doi:10.2514/1.18239.

[8] S. Cruchley, H. Evans, M. Taylor, An overview of the oxidation of Ni-based superalloys for turbine disc applications: surface condition, applied load and mechanical performance, Mater. High Temp. 33 (2016) 465–475. doi:10.1080/09603409.2016.1171952.

[9] A.G. Evans, J.W. Hutchinson, The mechanics of coating delamination in thermal gradients, Surf. Coatings Technol. 201 (2007) 7905–7916. doi:10.1016/j.surfcoat.2007.03.029.

[10] B.T. Richards, H.N.G. Wadley, Plasma spray deposition of tri-layer environmental barrier coatings, J. Eur. Ceram. Soc. 34 (2014) 3069–3083. doi:10.1016/j.jeurceramsoc.2014.04.027.

[11] E.J. Opila, N.S. Jacobson, D.L. Myers, E.H. Copland, Predicting oxide stability in high-temperature water vapor, JOM. 58 (2006) 22–28. doi:10.1007/s11837-006-0063-3.

[12] G.H. Meier, F.S. Pettit, K. Onal, Report: Interaction of Steam/Air MixturesWith Turbine Airfoil Alloys and Coatings, US Department of energy (2002).

[13] A. Sato, Y.-L. Chiu, R.C. Reed, Oxidation of nickel-based single-crystal superalloys for industrial gas turbine applications, Acta Mater. 59 (2011) 225–240.

doi:https://doi.org/10.1016/j.actamat.2010.09.027.

[14] H.T. Lin, M.K. Ferber, Mechanical reliability evaluation of silicon nitride ceramic components after exposure in industrial gas turbines, J. Eur. Ceram. Soc. 22 (2002) 2789–2797. doi:10.1016/S0955-2219(02)00146-2.

[15] N.S. Jacobson, Corrosion of Silicon-Based Ceramics in Combustion Environments, J. Am. Ceram. Soc. 76 (1993) 3–28. doi:10.1111/j.1151-2916.1993.tb03684.x.

[16] D.S. Fox, E.J. Opila, Q.G.N. Nguyen, D.L. Humphrey, S.M. Lewton, Paralinear oxidation of silicon nitride in a water-vapor/oxygen environment, J. Am. Ceram. Soc. 86 (2003) 1256–1261. doi:10.1111/j.1151-2916.2003.tb03461.x.

[17] E.J. Opila, R.C. Robinson, D.S. Fox, R.A. Wenglarz, M.K. Ferber, Additive Effects on Si_3N_4 Oxidation/Volatilization in Water Vapor, J. Am. Ceram. Soc. 86 (2003) 1262–1271. doi:10.1111/j.1151-2916.2003.tb03462.x.

[18] E.J. Opila, N.S. Jacobson, Corrosion of Ceramic Materials, in: Mater. Sci. Technol., Wiley-VCH Verlag GmbH & Co. KGaA, Weinheim, Germany, 2013. doi:10.1002/9783527603978.mst0409.

[19] H. Klemm, Silicon nitride for high-temperature applications, J. Am. Ceram. Soc. 93 (2010) 1501–1522. doi:10.1111/j.1551-2916.2010.03839.x.

[20] H. Klemm, Corrosion of silicon nitride materials in gas turbine environment, J. Eur. Ceram. Soc. 22 (2002) 2735–2740. doi:10.1016/S0955-2219(02)00143-7.

[21] F. Zok, Ceramic-matrix composites enable revolutionary gains in turbine engine efficiency, Am. Ceram. Soc. Bull. 95 (2016) 22–28.

[22] K.N. Lee, D.S. Fox, N.P. Bansal, Rare earth silicate environmental barrier coatings for SiC/SiC composites and Si_3N_4 ceramics, J. Eur. Ceram. Soc. 25 (2005) 1705–1715. doi:10.1016/j.jeurceramsoc.2004.12.013.

[23] M. Fritsch, H. Klemm, M. Herrmann, B. Schenk, Corrosion of selected ceramic materials in hot gas environment, J. Eur. Ceram. Soc. 26 (2006) 3557–3565. doi:10.1016/j.jeurceramsoc.2006.01.015.

[24] S. Ueno, D.D. Jayaseelan, T. Ohji, H.T. Lin, Corrosion and oxidation behavior of $ASiO_4$ (A=Ti, Zr and Hf) and silicon nitride with an $HfSiO_4$ environmental barrier coating, J. Ceram. Process. Res. 6 (2005) 81–84.

[25] S. Ueno, D.D. Jayaseelan, H. Kita, T. Ohji, H.T. Lin, Comparison of Water Vapor Corrosion Behaviors of $Ln_2Si_2O_7$ (Ln=Yb and Lu) and $ASiO_4$ (A=Ti, Zr and Hf) EBC's, Key Eng. Mater. 317–318 (2006) 557–560. doi:10.4028/www.scientific.net/kem.317-318.557.

[26] T. Bhatia, V. Vedula, H. Eaton, E. Sun, J. Holowczak, G. Linsey, Development and evaluation of environmental barrier coatings for Si-based ceramics, in: Proc. ASME Turbo Expo 2004, American Society of Mechanical Engineers, 2004: pp. 431–438. doi:10.1115/gt2004-54092.

[27] B.T. Richards, K.A. Young, F. De Francqueville, S. Sehr, M.R. Begley, H.N.G. Wadley, Response of ytterbium disilicate-silicon environmental barrier coatings to thermal cycling in water vapor, Acta Mater. 106 (2016) 1–14. doi:10.1016/j.actamat.2015.12.053.

[28] Y. Wang, J. Liu, First-principles investigation on the corrosion resistance of rare earth disilicates in water vapor, J. Eur. Ceram. Soc. 29 (2009) 2163–2167. doi:https://doi.org/10.1016/j.jeurceramsoc.2009.02.005.

[29] N. Maier, K.G. Nickel, G. Rixecker, High temperature water vapour corrosion of rare earth disilicates $(Y,Yb,Lu)_2Si_2O_7$ in the presence of $Al(OH)_3$ impurities, J. Eur. Ceram. Soc. 27 (2007) 2705–2713. doi:10.1016/j.jeurceramsoc.2006.09.013.

[30] G. Bocquillon, C. Chateau, C. Loriers, J. Loriers, Polymorphism under pressure of the disilicates of the heavier lanthanoids $Ln_2Si_2O_7$ (Ln = Tm, Yb, Lu), J. Solid State Chem. 20 (1977) 135–141. doi:https://doi.org/10.1016/0022-4596(77)90060-3.

[31] M. Ridley, E. Opila, Thermochemical stability and microstructural evolution of $Yb_2Si_2O_7$ in high-velocity high-temperature water vapor, J. Eur. Ceram. Soc. (2020). doi:10.1016/j.jeurceramsoc.2020.05.071.

[32] M. Wada, T. Matsudaira, N. Kawashima, S. Kitaoka, M. Takata, Mass transfer in polycrystalline ytterbium disilicate under oxygen potential gradients at high temperatures, Acta Mater. 135 (2017) 372–381. doi:10.1016/j.actamat.2017.06.029.

[33] T. Matsudaira, M. Wada, N. Kawashima, M. Takeuchi, D. Yokoe, T. Kato, M. Takata, S. Kitaoka, Mass transfer in polycrystalline ytterbium monosilicate under oxygen potential gradients at high temperatures, J. Eur. Ceram. Soc. (2020). doi:10.1016/j.jeurceramsoc.2020.07.045.

[34] E. Bakan, D. Marcano, D. Zhou, Y.J. Sohn, G. Mauer, R. Vaßen, $Yb_2Si_2O_7$ Environmental Barrier Coatings Deposited by Various Thermal Spray Techniques: A Preliminary Comparative Study, J. Therm. Spray Technol. 26 (2017) 1011–1024. doi:10.1007/s11666-017-0574-1.

[35] K.N. Lee, $Yb_2Si_2O_7$ Environmental barrier coatings with reduced bond coat oxidation rates via chemical modifications for long life, J. Am. Ceram. Soc. 102 (2019) 1507–1521. doi:10.1111/jace.15978.

[36] T. Bhatia, G. V. Srinivasan, S. V. Tulyani, R.A. Barth, V.R. Vedula, W.K. Tredway, Environmental barrier coatings for monolithic silicon nitride bond coat development, in: Proc. ASME Turbo Expo, American Society of Mechanical Engineers Digital Collection, 2007: pp. 281–288. doi:10.1115/GT2007-27685.

[37] N.P. Padture, Environmental degradation of high-temperature protective coatings for ceramic-matrix composites in gas-turbine engines, Npj Mater. Degrad. 3 (2019) 1–6. doi:10.1038/s41529-019-0075-4.

[38] E. Bakan, Y.J. Sohn, W. Kunz, H. Klemm, R. Vaßen, Effect of processing on high-velocity water vapor recession behavior of Yb-silicate environmental barrier coatings, J. Eur. Ceram. Soc. 39 (2019) 1507–1513. doi:10.1016/j.jeurceramsoc.2018.11.048.

[39] S.N. Basu, T. Kulkarni, H.Z. Wang, V.K. Sarin, Functionally graded chemical vapor deposited mullite environmental barrier coatings for Si-based ceramics, J. Eur. Ceram. Soc. 28 (2008) 437–445. doi:10.1016/j.jeurceramsoc.2007.03.007.

[40] J. Xu, V.K. Sarin, S. Dixit, S.N. Basu, Stability of interfaces in hybrid EBC/TBC coatings for Si-based ceramics in corrosive environments, Int. J. Refract. Met. Hard Mater. 49 (2015) 339–349. doi:10.1016/j.ijrmhm.2014.08.013.

[41] S. Ueno, D.D. Jayaseelan, T. Ohji, Development of Oxide-Based EBC for Silicon Nitride, Int. J. Appl. Ceram. Technol. 1 (2005) 362–373. doi:10.1111/j.1744-7402.2004.tb00187.x.

[42] P. Colombo, G. Mera, R. Riedel, G.D. Sorarù, Polymer-derived ceramics: 40 Years of research and innovation in advanced ceramics, J. Am. Ceram. Soc. 93 (2010) 1805–1837. doi:10.1111/j.1551-2916.2010.03876.x.

[43] G. Barroso, Q. Li, R.K. Bordia, G. Motz, Polymeric and ceramic silicon-based coatings-a review, J. Mater. Chem. A. (2019) 1936–1963. doi:10.1039/c8ta09054h.

[44] S. M. Attia, J. Wang, G. Wu, J. Shen, J. Ma, Review on Sol-Gel Derived Coatings: Process, Techniques and Optical Applications, J. Mater. Sci. Technol. 18 (2002) 211–218.

[45] S. Ueno, D. D. Jayaseelan, N. Kondo, T. Ohji, S. Kanzaki, H. T. Lin, Development of EBC for Silicon Nitride, Key Eng. Mater. 287 (2005) 449–456. doi:10.4028/www.scientific.net/kem.287.449.

[46] D. D. Jayaseelan, S. Ueno, T. Ohji, S. Kanzaki, Sol-gel synthesis and coating of nanocrystalline $Lu_2Si_2O_7$ on Si_3N_4 substrate, Mater. Chem. Phys. 84 (2004) 192–195. doi:10.1016/j.matchemphys.2003.11.028.

[47] G. Motz, T. Schmalz, S. Trassl, R. Kempe, Oxidation behavior of SiCN Materials, in: S. Bernard (Ed.), Des. Process. Prop. Ceram. Mater. from Preceramic Precursors, Nova Science Publishers Inc., 2011: pp. 15–36.

[48] P. Greil, Active-Filler-Controlled Pyrolysis of Preceramic Polymers, J. Am. Ceram. Soc. 78 (1995) 835–848. doi:10.1111/j.1151-2916.1995.tb08404.x.

[49] M. Günthner, K. Wang, R.K. Bordia, G. Motz, Conversion behaviour and resulting mechanical properties of polysilazane-based coatings, J. Eur. Ceram. Soc. 32 (2012) 1883–1892. doi:https://doi.org/10.1016/j.jeurceramsoc.2011.09.005.

[50] M. Günthner, T. Kraus, A. Dierdorf, D. Decker, W. Krenkel, G. Motz, Advanced coatings on the basis of Si(C)N precursors for protection of steel against oxidation, J. Eur. Ceram. Soc. 29 (2009) 2061–2068. doi:10.1016/j.jeurceramsoc.2008.11.013.

[51] M. Günthner, A. Schütz, U. Glatzel, K. Wang, R.K. Bordia, O. Greißl, W. Krenkel, G. Motz, High performance environmental barrier coatings, Part I: Passive filler loaded SiCN system for steel, J. Eur. Ceram. Soc. 31 (2011) 3003–3010. doi:10.1016/j.jeurceramsoc.2011.05.027.

[52] K. Wang, M. Günthner, G. Motz, R. K. Bordia, High performance environmental barrier coatings, Part II: Active filler loaded SiOC system for superalloys, J. Eur. Ceram. Soc. 31 (2011) 3011–3020. doi:10.1016/j.jeurceramsoc.2011.05.047.

[53] M. Günthner, T. Kraus, W. Krenkel, G. Motz, A. Dierdorf, D. Decker, Particle-Filled PHPS Silazane-Based Coatings on Steel, Int. J. Appl. Ceram. Technol. 6 (2009) 373–380. doi:10.1111/j.1744-7402.2008.02346.x.

[54] M. Lenz Leite, G. Barroso, M. Parchovianský, D. Galusek, E. Ionescu, W. Krenkel, G. Motz, Synthesis and characterization of yttrium and ytterbium silicates from their oxides and an oligosilazane by the PDC route for coating applications to protect Si_3N_4 in hot gas environments, J. Eur. Ceram. Soc. 37 (2017) 5177–5191. doi:10.1016/j.jeurceramsoc.2017.04.034.

[55] H. F. Chen, H. Klemm, Environmental Barrier Coatings for Silicon Nitride, in: Adv. Eng. Ceram. Compos., 2011: pp. 139–144. doi:10.4028/www.scientific.net/KEM.484.139.

[56] U. Bast, Thermal Shock and Cyclic Loading of Ceramic Parts in Stationary Gas Turbines, in: Therm. Shock Therm. Fatigue Behav. Adv. Ceram., Springer Netherlands, 1993: pp. 87–97. doi:10.1007/978-94-015-8200-1_8.

[57] M. Bentele, C.S. Lowthian, Thermal Shock Tests on Gas Turbine Materials: A Study of the Effects of Severe Temperature Fluctuations on Rotor Blades and Nozzle Segments, Aircr. Eng. Aerosp. Technol. 24 (1952) 32–38. doi:10.1108/eb032127.

[58] L. Witek, Stress Analysis of Turbine Components under Spin Rig Thermomechanical Conditition, Aviation. 8 (2004) 21–26. doi:10.3846/16487788.2004.9635885.

[59] Madhu P, Stress Analysis and Life Estimation of Gas Turbine Blisk for Different Materials of a Jet Engine, Int. J. Sci. Res. ISSN. 5 (2013). doi:10.21275/v5i6.NOV164440.

[60] W. Krenkel, Ceramic Matrix Composites: Fiber Reinforced Ceramics and their Applications, 2008. doi:10.1002/9783527622412.

[61] H. Salmang, H. Scholze, Keramik, 7th ed., Springer-Verlag Berlin Heidelberg, 2007. doi:10.1007/978-3-540-49469-0.

[62] N.P. Bansal, J. Lamon, Ceramic Matrix Composites: Materials, Modeling and Technology, 2014. doi:10.1002/9781118832998.

[63] M.J. Hoffmann, High-Temperature Properties of Si_3N_4 Ceramics, MRS Bull. 20 (1995) 28–32. doi:10.1557/S0883769400049186.

[64] Kostic, E, Sintering of silicon carbide in the presence of oxide additives, Powder Met. Int.; (Germany, Fed. Repub. Of). 20:6 (1988).

[65] F.K. Van Dijen, E. Mayer, Liquid phase sintering of silicon carbide, J. Eur. Ceram. Soc. 16 (1996) 413–420. doi:10.1016/0955-2219(95)00129-8.

[66] R. Kochendörfer, W. Krenkel, CMC-intake ramp for hypersonic propulsion systems, Ceram. Trans. 57 (1995) 13–22. https://elib.dlr.de/40307/.

[67] L.U.J. Thomas-Ogbuji, Silicon-Based Ceramic-Matrix Composites for Advanced Turbine Engines: Some Degradation Issues, Cleveland, OH 44122, USA, 2000.

[68] Y. Gowayed, G. Ojard, R. Miller, U. Santhosh, J. Ahmad, R. John, Correlation of elastic properties of melt infiltrated SiC/SiC composites to in situ properties of constituent phases, Compos. Sci. Technol. 70 (2010) 435–441. doi:https://doi.org/10.1016/j.compscitech.2009.11.016.

[69] N.P. Bansal, Handbook of Ceramic Composites, Springer US, 2005. doi:10.1007/b104068.

[70] D.W. Richerson, Ceramic Components in Gas Turbine Engines: Why Has It Taken So Long?, in: E. Lara-Curzio, M.J. Readey (Eds.), 28th Int. Conf. Adv. Ceram. Compos. A Ceram. Eng. Sci. Proc., John Wiley & Sons, Ltd, 2008: pp. 3–32. doi:10.1002/9780470291184.ch1.

[71] B.E. Deal, A.S. Grove, General Relationship for the Thermal Oxidation of Silicon, J. Appl. Phys. 36 (1965) 3770–3778. doi:10.1063/1.1713945.

[72] K. H. Stern, Oxidation of Silicon, Silicon Carbide (SiC) and Silicon Nitride (Si_3N_4), Naval Research Laboratory, Washington, DC, USA, 1986.

[73] A.J. Kiehle, L.K. Heung, P.J. Gielisse, T.J. Rockett, Oxidation Behavior of Hot-Pressed Si_3N_4, J. Am. Ceram. Soc. 58 (1975) 17–20. doi:10.1111/j.1151-2916.1975.tb18972.x.

[74] D. Cubicciotti, K.H. Lau, Kinetics of Oxidation of Yttria Hot-Pressed Silicon Nitride, J. Electrochem. Soc. 126 (1979) 1723–1728. doi:10.1149/1.2128785.

[75] D. Cubicciotti, K.H. Lau, Kinetics of Oxidation of Hot-Pressed Silicon Nitride Containing Magnesia, J. Am. Ceram. Soc. 61 (1978) 512–517. doi:10.1111/j.1151-2916.1978.tb16130.x.

[76] H. Klemm, C. Taut, G. Wötting, Long-term stability of nonoxide ceramics in an oxidative environment at 1500 °C, J. Eur. Ceram. Soc. 23 (2003) 619–627. doi:10.1016/S0955-2219(02)00267-4.

[77] M.K. Cinibulk, G. Thomas, S.M. Johnson, Fabrication and Secondary-Phase Crystallization of Rare-Earth Disilicate–Silicon Nitride Ceramics, J. Am. Ceram. Soc. 75 (1992) 2037–2043. doi:10.1111/j.1151-2916.1992.tb04462.x.

[78] M.K. Cinibulk, Design, Microstructure and High-Temperature Behavior of Silicon Nitride Sintered with Rare-Earth Oxides, University of California, 1991. https://escholarship.org/uc/item/5vv1f41b.

[79] S.C. Singhal, Effect of Water Vapor on the Oxidation of Hot-Pressed Silicon Nitride and Silicon Carbide, J. Am. Ceram. Soc. 59 (1976) 81–82. doi:10.1111/j.1151-2916.1976.tb09398.x.

[80] M. Maeda, K. Nakamura, T. Ohkubo, Oxidation of silicon nitride in a wet atmosphere, J. Mater. Sci. 24 (1989) 2120–2126. doi:10.1007/BF02385430.

[81] M. Maeda, K. Nakamura, M. Yamada, Oxidation resistance of silicon nitride ceramics with various additives, J. Mater. Sci. 25 (1990) 3790–3794. doi:10.1007/BF00575420.

[82] E. Courcot, F. Rebillat, F. Teyssandier, C. Louchet-Pouillerie, Stability of rare earth oxides in a moist environment at elevated temperatures-Experimental and thermodynamic studies. Part II: Comparison of the rare earth oxides, J. Eur. Ceram. Soc. 30 (2010) 1911–1917. doi:10.1016/j.jeurceramsoc.2010.02.012.

[83] M. Fritsch, Heißgaskorrosion keramischer Werkstoffe in H_2O-haltigen Rauchgasatmosphären, Monografie, Technische Universität Dresden, 2007.

[84] T.H. Nielsen, M.H. Leipold, Thermal Expansion of Yttrium Oxide and of Magnesium Oxide with Yttrium Oxide, J. Am. Ceram. Soc. 47 (1964) 256–256. doi:10.1111/j.1151-2916.1964.tb14408.x.

[85] Z. Sun, M. Li, Y. Zhou, Recent progress on synthesis, multi-scale structure, and properties of Y–Si–O oxides, Int. Mater. Rev. 59 (2014) 357–383. doi:10.1179/1743280414Y.0000000033.

[86] K.N. Lee, Current status of environmental barrier coatings for Si-based ceramics, Surf. Coatings Technol. 133–134 (2000) 1–7. doi:10.1016/S0257-8972(00)00889-6.

[87] K.N. Lee, D.S. Fox, J.I. Eldridge, D. Zhu, R.C. Robinson, N.P. Bansal, R.A. Miller, Upper Temperature Limit of Environmental Barrier Coatings Based on Mullite and BSAS, J. Am. Ceram. Soc. 86 (2003) 1299–1306. doi:10.1111/j.1151-2916.2003.tb03466.x.

[88] K.N. Lee, J.I. Eldridge, R.C. Robinson, Residual Stresses and Their Effects on the Durability of Environmental Barrier Coatings for SiC Ceramics, J. Am. Ceram. Soc. 88 (2005) 3483–3488. doi:10.1111/j.1551-2916.2005.00640.x.

[89] H.E. Eaton, G.D. Linsey, K.L. More, J.B. Kimmei, J.R. Price, N. Miriyala, EBC protection of SiC/SiC composites in the gas turbine combustion environment, in: Proc. ASME Turbo Expo, American Society of Mechanical Engineers (ASME), 2000. doi:10.1115/2000-GT-0631.

[90] K.L. More, P.F. Tortorelli, L.R. Walker, J.B. Kimmel, N. Miriyala, J.R. Price, H.E. Eaton, E.Y. Sun, G.D. Linsey, Evaluating environmental barrier coatings on ceramic matrix composites after engine and laboratory exposures, in: Am. Soc. Mech. Eng. Int. Gas Turbine Institute, Turbo Expo IGTI, American Society of Mechanical Engineers Digital Collection, 2002: pp. 155–162. doi:10.1115/GT2002-30630.

[91] H. Yang, Y. Yang, X. Cao, X. Huang, Y. Li, Thermal shock resistance and bonding strength of tri-layer Yb_2SiO_5/mullite/Si coating on SiC_f/SiC composites, Ceram. Int. 46 (2020) 27292–27298. doi:10.1016/j.ceramint.2020.07.214.

[92] S. Ueno, T. Ohji, H.T. Lin, Recession behavior of a silicon nitride with multi-layered environmental barrier coating system, Ceram. Int. 33 (2007) 859–862.

doi:10.1016/j.ceramint.2006.01.012.

[93] P. Colombo, R. Riedel, G.D. Sorarù, H.-J. Kleebe, Polymer derived ceramics: from nanostructure to applications, DEStech Publications, Inc, Lancaster, PA, USA, 2010.

[94] R. Chavez, E. Ionescu, C. Balan, C. Fasel, R. Riedel, Effect of ambient atmosphere on crosslinking of polysilazanes, J. Appl. Polym. Sci. 119 (2010) 794–802.

[95] E. Kroke, Y.-L. Li, C. Konetschny, E. Lecomte, C. Fasel, R. Riedel, Silazane derived ceramics and related materials, Mater. Sci. Eng., R. 26 (2000) 97–199.

[96] G. Ziegler, H.J. Kleebe, G. Motz, H. Müller, S. Traßl, W. Weibelzahl, Synthesis, microstructure and properties of SiCN ceramics prepared from tailored polymers, Mater. Chem. Phys. 61 (1999) 55–63. doi:10.1016/S0254-0584(99)00114-5.

[97] O. Flores, T. Schmalz, W. Krenkel, L. Heymann, G. Motz, Selective cross-linking of oligosilazanes to tailored meltable polysilazanes for the processing of ceramic SiCN fibres, J. Mater. Chem. A. 1 (2013) 15406. doi:10.1039/c3ta13254d.

[98] Y. Iwamoto, W. Völger, E. Kroke, R. Riedel, T. Saitou, K. Matsunaga, Crystallization behavior of amorphous silicon carbonitride ceramics derived from organometallic precursors, J. Am. Ceram. Soc. 84 (2001) 2170–2178.

[99] O. Goerke, E. Feike, T. Heine, A. Trampert, H. Schubert, Ceramic coatings processed by spraying of siloxane precursors (polymer-spraying), J. Eur. Ceram. Soc. 24 (2004) 2141–2147. doi:https://doi.org/10.1016/S0955-2219(03)00362-5.

[100] J. Bill, D. Heimann, Polymer-Derived Ceramic Coatings on C/C-SiC Composites, J. Eur. Ceram. Soc. 16 (1996) 1115–1120. doi:10.1016/0955-2219(96)00025-8.

[101] V. Bakumov, K. Gueinzius, C. Hermann, M. Schwarz, E. Kroke, Polysilazane-derived antibacterial silver-ceramic nanocomposites, J. Eur. Ceram. Soc. (2007) 3287–3292. doi:10.1016/j.jeurceramsoc.2007.01.004.

[102] A. Schütz, M. Günthner, G. Motz, O. Greißl, U. Glatzel, High temperature (salt melt) corrosion tests with ceramic-coated steel, Mater. Chem. Phys. 159 (2015) 10–18. doi:https://doi.org/10.1016/j.matchemphys.2015.03.023.

[103] G.S. Barroso, W. Krenkel, G. Motz, Low thermal conductivity coating system for application up to 1000 °C by simple PDC processing with active and passive fillers, J. Eur. Ceram. Soc. (2015) 3339–3348. doi:10.1016/j.jeurceramsoc.2015.02.006.

[104] G. Barroso, T. Kraus, U. Degenhardt, M. Scheffler, G. Motz, Functional Coatings Based on Preceramic Polymers, Adv. Eng. Mater. 18 (2016) 746–753. doi:10.1002/adem.201500600.

[105] K. Tangermann-Gerk, G. Barroso, B. Weisenseel, P. Greil, T. Fey, M. Schmidt, G. Motz, Laser pyrolysis of an organosilazane-based glass/ZrO_2 composite coating system, Mater. Des. 109 (2016) 644–651. doi:10.1016/j.matdes.2016.07.102.

[106] A. Horcher, K. Tangermann-Gerk, G. Barroso, M. Schmidt, G. Motz, Laser and furnace

pyrolyzed organosilazane-based glass/ZrO_2 composite coating system - A comparison, J. Eur. Ceram. Soc. 40 (2020) 2642–2651. doi:10.1016/j.jeurceramsoc.2019.10.040.

[107] M. Aparicio, A. Durán, Yttrium silicate coatings for oxidation protection of carbon-silicon carbide composites, J. Am. Ceram. Soc. 83 (2000) 1351–1355. doi:10.1111/j.1151-2916.2000.tb01392.x.

[108] N. Al Nasiri, N. Patra, D. Horlait, D.D. Jayaseelan, W.E. Lee, Thermal Properties of Rare-Earth Monosilicates for EBC on Si-Based Ceramic Composites, J. Am. Ceram. Soc. 99 (2016) 589–596. doi:10.1111/jace.13982.

[109] N. Maier, G. Rixecker, K.G. Nickel, Formation and stability of Gd, Y, Yb and Lu disilicates and their solid solutions, J. Solid State Chem. 179 (2006) 1630–1635. doi:10.1016/j.jssc.2006.02.019.

[110] N. Maier, K.G. Nickel, G. Rixecker, High temperature water vapour corrosion of rare earth disilicates $(Y,Yb,Lu)_2Si_2O_7$ in the presence of $Al(OH)_3$ impurities, J. Eur. Ceram. Soc. 27 (2007) 2705–2713. doi:https://doi.org/10.1016/j.jeurceramsoc.2006.09.013.

[111] S.-B. Wang, Y.-R. Lu, Y.-X. Chen, Synthesis of Single-Phase β-$Yb_2Si_2O_7$ and Properties of Its Sintered Bulk, Int. J. Appl. Ceram. Technol. 12 (2015) 1140–1147. doi:10.1111/ijac.12353.

[112] E. Bernardo, G. Parcianello, P. Colombo, Novel synthesis and applications of yttrium silicates from a silicone resin containing oxide nano-particle fillers, Ceram. Int. 38 (2012) 5469–5474. doi:10.1016/j.ceramint.2012.03.059.

[113] J. Liu, L. Zhang, F. Hu, J. Yang, L. Cheng, Y. Wang, Polymer-derived yttrium silicate coatings on 2D C/SiC composites, J. Eur. Ceram. Soc. 33 (2013) 433–439. doi:10.1016/j.jeurceramsoc.2012.08.032.

[114] L. An, Y. Wang, L. Bharadwaj, L. Zhang, Y. Fan, D. Jiang, Y. Sohn, V.H. Desai, J. Kapat, L.C. Chow, Silicoaluminum Carbonitride with Anomalously High Resistance to Oxidation and Hot Corrosion, Adv. Eng. Mater. 6 (2004) 337–340. doi:10.1002/adem.200400010.

[115] M. Seifert, M. Lenz Leite, G. Motz, Formation of Mg-silicates by reaction of perhydropolysilazane with MgO during pyrolysis in air, J. Ceram. Soc. Japan. 124 (2016) 1003–1005. doi:10.2109/jcersj2.16026.

[116] S. Stecura, W.J. Campbell, Thermal expansion and phase inversion of rare-earth oxides, 1960.

[117] FCT Si_3N_4 Sondewerkstoffe, (2019). https://fcti.de/images/PDF_zum_Downloaden/FCTSi3N4-Sonderwerkstoffe2019_10de.pdf (accessed 20 July 2021).

[118] F. Riley, Progress in Nitrogen Ceramics, Springer, Dordrecht, The Hague, 1983. doi:10.1007/978-94-009-6851-6.

[119] Y.G. Gogotsi, V.A. Lavrenko, Corrosion of High-Performance Ceramics, Springer-

Verlag Berlin Heidelberg, 1992. doi:10.1007/978-3-642-77390-7.

[120] A. Morlier, S. Cros, J.-P. Garandet, N. Alberola, Thin gas-barrier silica layers from perhydropolysilazane obtained through low temperature curings: A comparative study, Thin Solid Films. 524 (2012) 62–66. doi:https://doi.org/10.1016/j.tsf.2012.09.065.

[121] T. Sato, K. Haryu, T. Endo, M. Shimada, High-temperature oxidation of silicon nitride-based ceramics by water vapour, J. Mater. Sci. 22 (1987) 2635–2640. doi:10.1007/BF01082156.

[122] C.A. Schneider, W.S. Rasband, K.W. Eliceiri, NIH Image to ImageJ: 25 years of image analysis, Nat. Methods. 9 (2012) 671–675. doi:10.1038/nmeth.2089.

[123] N.F. Mott, S. Rigo, F. Rochet, A.M. Stoneham, Oxidation of silicon, Philos. Mag. B. 60 (1989) 189–212. doi:10.1080/13642818908211190.

[124] M. Németh-Sallay, Barna, Lőrinczy, I.C. Szép, Induced crystallization of amorphous silicon oxide, Phys. Status Solidi. 43 (1977) 1–3. doi:10.1002/pssa.2210430249.

[125] N.F. Mott, On the oxidation of silicon, Philos. Mag. B Phys. Condens. Matter; Stat. Mech. Electron. Opt. Magn. Prop. 55 (1987) 117–129. doi:10.1080/13642818708211199.

[126] T.L. Christiansen, O. Hansen, J.A. Jensen, E.V. Thomsen, Thermal Oxidation of Structured Silicon Dioxide, ECS J. Solid State Sci. Technol. 3 (2014) N63–N68. doi:10.1149/2.003405jss.

[127] A.G. Revesz, B.J. Mrstik, H.L. Hughes, Thermal Oxidation of Silicon, in: Phys. Technol. Amorph. SiO_2, Springer US, Boston, MA, 1988: pp. 297–306. doi:10.1007/978-1-4613-1031-0_41.

[128] T. Miyazaki, S. Usami, R. Inoue, Y. Kogo, Y. Arai, Oxidation behavior of ytterbium silicide, Ceram. Int. 45 (2019) 9560–9566. doi:https://doi.org/10.1016/j.ceramint.2018.10.024.

[129] E. Garcia, H. Lee, S. Sampath, Phase and microstructure evolution in plasma sprayed $Yb_2Si_2O_7$ coatings, J. Eur. Ceram. Soc. 39 (2019) 1477–1486. doi:10.1016/j.jeurceramsoc.2018.11.018.

[130] B.T. Richards, H. Zhao, H.N.G. Wadley, Structure, composition, and defect control during plasma spray deposition of ytterbium silicate coatings, J. Mater. Sci. 50 (2015) 7939–7957. doi:10.1007/s10853-015-9358-5.

[131] Y.-C. Zhou, C. Zhao, F. Wang, Y.-J. Sun, L.-Y. Zheng, X.-H. Wang, Theoretical Prediction and Experimental Investigation on the Thermal and Mechanical Properties of Bulk β-$Yb_2Si_2O_7$, J. Am. Ceram. Soc. 96 (2013) 3891–3900. doi:10.1111/jace.12618.

[132] D.R. Clarke, Thermodynamic Mechanism for cation diffusion through intergranular phase: application to environmental reactions with nitrogen ceramics, in: F.L. Riley (Ed.), Prog. Nitrogen Ceram., Springer, Dordrecht, The Hague, 1983: pp. 421–426. doi:10.1007/978-94-009-6851-6.

[133] F. Monteverde, A. Bellosi, High oxidation resistance of hot pressed silicon nitride containing yttria and lanthania, J. Eur. Ceram. Soc. 18 (1998) 2313–2321. doi:10.1016/S0955-2219(98)00237-4.

[134] G. Zhang, Y. Xu, D. Xu, D. Wang, Y. Xue, W. Su, Pressure-induced crystallization of amorphous SiO_2 with silicon-hydroxy group and the quick synthesis of coesite under lower temperature, in: High Press. Res., Taylor & Francis , 2008: pp. 641–650. doi:10.1080/08957950802510091.

[135] S. Zhang, S. Tie, F. Zhang, Cristobalite formation from the thermal treatment of amorphous silica fume recovered from the metallurgical silicon industry, Micro Nano Lett. 13 (2018) 1465–1468. doi:10.1049/mnl.2018.5167.

[136] P. Bettermann, F. Liebau, The transformation of amorphous silica to crystalline silica under hydrothermal conditions, Contrib. to Mineral. Petrol. 53 (1975) 25–36. doi:10.1007/BF00402452.

[137] D.H. Olson, J.T. Gaskins, J.A. Tomko, E.J. Opila, R.A. Golden, G.J.K. Harrington, A.L. Chamberlain, P.E. Hopkins, Local thermal conductivity measurements to determine the fraction of α-cristobalite in thermally grown oxides for aerospace applications, Scr. Mater. 177 (2020) 214–217. doi:10.1016/j.scriptamat.2019.10.027.

[138] H.-J. Choi, J.-G. Lee, Y.-W. Kim, High temperature strength and oxidation behaviour of hot-pressed silicon nitride-disilicate ceramics, J. Mater. Sci. 32 (1997) 1937–1942. doi:10.1023/A:1018581529225.

[139] J.E. Shelby, J.T. Kohli, Rare-Earth Aluminosilicate Glasses, J. Am. Ceram. Soc. 73 (1990) 39–42. doi:10.1111/j.1151-2916.1990.tb05087.x.

[140] G.-H. Zhang, K.-C. Chou, K. Mills, Modelling Viscosities of $CaO/MgO/Al_2O_3/SiO_2$ Molten Slags, ISIJ Int. 52 (2012) 355–362. doi:10.2355/isijinternational.52.355.

[141] M. Futakawa, R.W. Steinbrech, Viscosity of Amorphous Oxide Scales on SiSiC at Elevated Temperatures, J. Am. Ceram. Soc. 81 (1998) 1819–1823. doi:10.1111/j.1151-2916.1998.tb02552.x.

[142] V. Pareek, D.A. Shores, Oxidation of Silicon Carbide in Environments Containing Potassium Salt Vapor, J. Am. Ceram. Soc. 74 (1991) 556–563. doi:10.1111/j.1151-2916.1991.tb04059.x.

[143] Z. Zhang, Y. Xiao, J. Voncken, Y. Yang, R. Boom, N. Wang, Z. Zou, Phase Equilibria in the $Na_2O–CaO–SiO_2$ System, J. Am. Ceram. Soc. 94 (2011) 3088–3093. doi:10.1111/j.1551-2916.2011.04442.x.

[144] H. Mao, O. Fabrichnaya, M. Selleby, B. Sundman, Thermodynamic assessment of the $MgO–Al_2O_3–SiO_2$ system, J. Mater. Res. 20 (2005) 975–986. doi:DOI: 10.1557/JMR.2005.0123.

[145] D. Chen, A. Pegler, M. Dorfman, Environmental barrier coatings using low pressure plasma spray process, J. Am. Ceram. Soc. 103 (2020) 4840–4845. doi:10.1111/jace.17199.

[146] J. Richard, S.L. Cid, J. Rouquette, A. van der Lee, S. Bernard, J. Haines, Pressure-induced insertion of ammonia borane in the siliceous zeolite, silicalite-1F, J. Phys. Chem. C. 120 (2016) 9334–9340.

[147] S.J. Bull, E.G. Berasetegui, An overview of the potential of quantitative coating adhesion measurement by scratch testing, Tribol. Int. 39 (2006) 99–114. doi:10.1016/j.triboint.2005.04.013.

[148] P.J. Burnett, D.S. Rickerby, The relationship between hardness and scratch adhesion, Thin Solid Films. 154 (1987) 403–416.

[149] J.B. Quinn, G.D. Quinn, A practical and systematic review of Weibull statistics for reporting strengths of dental materials, Dent. Mater. 26 (2010) 135–147. doi:10.1016/j.dental.2009.09.006.

[150] D.P.H. Hasselman, Unified Theory of Thermal Shock Fracture Initiation and Crack Propagation in Brittle Ceramics, J. Am. Ceram. Soc. 52 (1969) 600–604. doi:10.1111/j.1151-2916.1969.tb15848.x.

[151] D.P.H. Hasselman, Approximate Theory of Thermal Stress Resistance of Brittle Ceramics Involving Creep, J. Am. Ceram. Soc. 50 (1967) 454–457. doi:10.1111/j.1151-2916.1967.tb15160.x.

[152] M. Yoshimura, J. Kase, S. Somiya, Oxidation of Si_3N_4 and SiC by high temperature - high pressure water vapor, in: Hydrothermal React. Mater. Sci. Eng., Verlag Deutsche Keramische Gesellschaft, 1986: pp. 450–457. doi:10.1007/978-94-009-0743-0_78.

[153] M. Yoshimura, J. Kase, S. Sōmiya, Oxidation of SiC powder by high-temperature, high-pressure H_2O, in: Hydrothermal React. Mater. Sci. Eng., Springer Netherlands, 1989: pp. 390–393. doi:10.1007/978-94-009-0743-0_65.

[154] P.F. Tortorelli, K.L. More, Effects of high water-vapor pressure on oxidation of silicon carbide at 1200 °C, J. Am. Ceram. Soc. (2003). doi:10.1111/j.1151-2916.2003.tb03460.x.

[155] T. Cheng, P.F. Tortorelli, Silicon carbide oxidation in high-pressure steam, J. Am. Ceram. Soc. (2013). doi:10.1111/jace.12328.

[156] T. Nishimura, M. Mitomo, H. Suematsu, High temperature strength of silicon nitride ceramics with ytterbium silicon oxynitride, J. Mater. Res. 12 (1997) 203–209. doi:10.1557/JMR.1997.0027.

[157] M. Zaheer, T. Schmalz, G. Motz, R. Kempe, Polymer derived non-oxide ceramics modified with late transition metals, Chem. Soc. Rev. 41 (2012) 5102. doi:10.1039/c2cs15326b.

[158] M. Zaheer, G. Motz, R. Kempe, The generation of palladium silicide nanoalloy particles in a SiCN matrix and their catalytic applications, J. Mater. Chem. 21 (2011) 18825–18831. doi:10.1039/c1jm13665h.

[159] G. Glatz, T. Schmalz, T. Kraus, F. Haarmann, G. Motz, R. Kempe, Copper-containing

SiCN precursor ceramics (Cu@SiCN) as selective hydrocarbon oxidation catalysts using air as an oxidant, Chem. - A Eur. J. 16 (2010) 4231–4238. doi:10.1002/chem.200902836.

[160] E. Ionescu, B. Papendorf, H.J. Kleebe, H. Breitzke, K. Nonnenmacher, G. Buntkowsky, R. Riedel, Phase separation of a hafnium alkoxide-modified polysilazane upon polymer-to-ceramic transformation-A case study, J. Eur. Ceram. Soc. 32 (2012) 1873–1881. doi:10.1016/j.jeurceramsoc.2011.09.003.

[161] D. Forberg, T. Schwob, M. Zaheer, M. Friedrich, N. Miyajima, R. Kempe, Single-catalyst high-weight% hydrogen storage in an N-heterocycle synthesized from lignin hydrogenolysis products and ammonia, Nat. Commun. 7 (2016) 13201. doi:10.1038/ncomms13201.

[162] J. Yuan, M. Galetz, X.G. Luan, C. Fasel, R. Riedel, E. Ionescu, High-temperature oxidation behavior of polymer-derived SiHfBCN ceramic nanocomposites, J. Eur. Ceram. Soc. 36 (2016) 3021–3028. doi:https://doi.org/10.1016/j.jeurceramsoc.2015.12.006.

[163] E. Ionescu, S. Bernard, R. Lucas, P. Kroll, S. Ushakov, A. Navrotsky, R. Riedel, Polymer-Derived Ultra-High Temperature Ceramics (UHTCs) and Related Materials, Adv. Eng. Mater. 21 (2019) 1900269. doi:10.1002/adem.201900269.

[164] M. Lenz Leite, A. Viard, D. Galusek, G. Motz, In Situ Generated β-$Yb_2Si_2O_7$ Containing Coatings for Steel Protection in Extreme Combustion Environments, Adv. Mater. Interfaces. 8 (2021). doi:10.1002/admi.202170075.

[165] J. Vogt, A. Nöth, Slurry-Based Environmental Barrier Coatings for Silicon Carbide and its Composites - a Straightforward approach, Ceram. Appl. (2020) 32–36.

[166] U. Kolitsch, H.J. Seifert, T. Ludwig, F. Aldinger, Phase equilibria and crystal chemistry in the Y_2O_3-Al_2O_3-SiO_2 system, J. Mater. Res. 14 (1999) 447–455. doi:10.1557/JMR.1999.0064.

[167] Y. Murakami, H. Yamamoto, Phase equilibria and properties of glasses in the Al_2O_3-Yb_2O_3-SiO_2 system, J. Ceram. Soc. Japan 101 (1993) 1101-1106.

9 Acknowledgements

This work was accomplished at the Chair of Ceramic Materials (CME) of the University of Bayreuth in the frame of the ZIM cooperation program project number ZF4109902AG7 (Federal Ministry of Germany for Economic Affairs and Energy, BMWi) and of the project 453000562 (MO 851/20, Deutsche Forschungsgemeinschaft, DFG).

I thank Professor Dr.-Ing. Walter Krenkel for the opportunity to develop my thesis at CME and for the support and orientation until the completion of my work. To Professor Dr.-Ing. Ralf Moos and to Professor Dr.-Ing. Uwe Glatzel for kindly accepting the invitation to be a part of the examination committee.

My gratitude to Dr. rer. nat. habil. Günter Motz who offered me support along the way, for the mentoring and supervision of my work, besides the freedom to follow my own ideas, and for the many fruitful discussions.

I extend my gratitude to my former colleagues Alexander Horcher, Jan-Felix Wendel, Jakob Denk, Dr. Antoine Viard, Dr.-Ing. Gilvan Barroso and Dr. Plinio Furtat for the teamwork and support. To Sven Scheler and Walter Müller who not only contributed to my formation and helped me along the way, but also created a joyful working atmosphere. To Dr.-Ing. Ulrich Degenhardt for the cooperative work and fruitful discussions.

To my family and specially my parents who recognized the value of education, always stood by my side, and encouraged me to pursue my career. To my dearest friends for the support and motivation.

My gratitude.

10 Annexes

10.1 List of symbols and abbreviation

ΔM	Mass loss
$\Delta V/V_o$	Volume change
°C	Degrees Celsius
µ	Micro (10-6)
∅	Diameter
APS	Air plasma spray
AS800	Turbine grade silicon nitride containing $La_2O_3/Y_2O_3/SrO$ (Honeywell Aerospace, USA)
BSAS	$1\text{-}xBaO\text{-}xSrO\text{-}Al_2O_3\text{-}2SiO_2$, $0 \leq x \leq 1$
C/C-SiC	Carbon fiber reinforced silicon carbide composite
C/SiC	Carbon fiber reinforced silicon carbide
CMC	Ceramic matrix composites
CTE	Coefficient of thermal expansion
CVD	Chemical vapor deposition
CVI	Chemical vapor infiltration
DBE	Di-n-butyl-ether
DCP	Dicumyl peroxide
EBC	Environmental barrier coating
EDS	Energy dispersive spectroscopy
FOD	Foreign object damage
g	Gram
GPSN	Gas pressure sintered silicon nitride
h	Hour
HIPSiC	Hot isostatic pressed silicon carbide
HIPSN	Hot isostatic pressed silicon nitride
HPBR	High-pressure burner rig
HPSiC	Hot pressed silicon carbide
HPSN	Hot pressed silicon nitride
ICP	Inductively coupled plasma mass spectrometry
K	Kelvin
k	Kilo
LPSiC	Liquid phase sintered silicon carbide
M	Mega
m	Meter

m	Weibull modulus
M'	Mass of the Yb(2:1) monoliths after pyrolysis
min	Minutes
Mw	Molar mass
p	Pressure
P	Probability of failure
Pa	Pascal (pressure)
PDC	Polymer derived ceramics
p_{H2O}	Water vapor partial pressure
PHPS	Perhydropolysilazane
p_{O2}	Oxygen partial pressure
PVD	Physical vapor deposition
RBSN	Reaction bonded silicon nitride
RE	Rare-earth
$RE(OH)_3$	Rare-earth hydroxide
RE_2O_3	Rare-earth oxide
$RE_2Si_2O_7$	Rare-earth disilicate
RE_2SiO_5	Rare-earth monosilicate
RE^{3+}	Rare-earth cation
REOOH	Rare-earth hydroxide
RT	Room temperature
s	Second
SEM	Scanning electron microscopy
SiB	Polyborosilanes
SiC	Polycarbosilanes
SiC/SiC	Fiber reinforced silicon carbide matrix composites
SiCN	Polycarbosilazanes
SiCO	Polycarbosiloxanes
SiN	Polysilazanes
SiO	Polysiloxanes
SiSiC	Silicon infiltrated silicon carbide
SN282	Turbine grade silicon nitride containing Lu_2O_3 as sintering additive (Kyocera Corporation, Japan)
SNPU	Sinteradditiv-armes Si_3N_4 (selected substrate)
SSN	Pressureles sintered silicon nitride
T	Temperature

t	Time
TAK	Thermischer Ausdehnungskoeffizient
TBC	Thermal barrier coating
TEOS	Tetraethoxysilane
TGA	Thermogravimetric analysis
TGO	Thermally grown oxide
T_{melt}	Melting temperature
v	velocity
V'	Volume of Yb(2:1) before pyrolysis
VHe	Volume measured by He-pycnometry
Vo	Volume of Yb(2:1) after pyrolysis
W	Watt
XRD	X-ray diffractometry
YAG	Yttrium aluminium garnet
YAS	Yttrium aluminosilicate
YbAS	Ytterbium aluminosilicate
YSZ	Yttria-stabilized-zirconia
ρ	Density
σ_o	Characteristic strength
σ	Flexural strength
φ	Open porosity content

10.2 List of chemical compounds

$Al(OH)_3$	Aluminium hydroxide
Al_2O_3	Aluminium oxide or alumina
AlB_4	Aluminium boride
B_4C	Boron carbide
BaO	Barium oxide
BN	Boron nitride
$Ca_2MgSi_2O_7$	Calcium magnesium oxide
CaO	Calcium oxide
CH_4	Methane
CO	Carbon monoxide
CO_2	Carbon dioxide
Er_2SiO_5	Erbium monosilicate

H_2	Hydrogen gas
H_2O	Water
$HfSiO_4$	Hafnium silicate
La_2O_3	Lanthanium oxide
$Lu(NO_3)_3$	Lutetium nitrate
Lu_2O_3	Lutetium oxide
$Lu_2Si_2O_7$	Lutetium disilicate
Lu_2SiO_5	Lutetium monosilicate
Mg_2SiO_4	Magnesium silicate
MgO	Magnesium oxide
$MgSiO_3$	Magnesium silicate
$MoSi_2$	Molibdenium silicide
N_2	Nitrogen gas
NH_3	Ammonia
NOx	Nitrogen oxides
Sc_2O_3	Scandium oxide
$Sc_2Si_2O_7$	Scandium disilicate
Sc_2SiO_5	Scandium monosilicate
Si	Silicon
$Si(OH)_4$	Silicon hydroxide
Si_3N_4	Silicon nitride
SiB_6	Silicon boride
SiC	Silicon carbide
SiO	Silicon monoxide
SiO_2	Silicon dioxide or silica
SrO	Stroncium oxide
$TiSiO_4$	Titanium silicate
Y_2O_3	Yttrium oxide
$Y_2Si_2O_7$	Yttrium disilicate
Y_2SiO_5	Yttrium monosilicate
$Y_3Al_5O_{12}$	Yttrium-aluminium garnet
Yb_2O_3	Ytterbium oxide or ytterbia
$Yb_2Si_2O_7$	Ytterbium disilicate
Yb_2SiO_5	Ytterbium monosilicate
Yb_3C_4	Ytterbium carbide

$Yb_4Si_2N_2O_7$	Ytterbium oxy-nitride silicate
Yb_4Si_7	Ytterbium silicide
$Yb_6Si_{11}N_{20}O$	Ytterbium oxy-nitride silicate
ZrO_2	Zirconium oxide
$ZrSi_2$	Zirconium silicide
$ZrSiO_4$	Zirconium silicate

10.3 List of figures

10.4 List of tables

10.5 Publications

Lenz Leite, M.; Degenhardt, U..; Krenkel, W.; Schafföner, S.; Motz, G. **In Situ Generated $Yb_2Si_2O_7$ Environmental Barrier Coatings for Protection of Ceramic Components in the Next Generation of Gas Turbines**. *Adv. Mater. Interfaces*, **2022**, 2102255. https://doi.org/10.1002/admi.202102255

Lenz Leite, M.; Viard, A.; Galusek, D.; Motz, G. **In-Situ Generated β-$Yb_2Si_2O_7$ Containing Coatings for Steel Protection in Extreme Combustion Environments**. *Adv. Mater. Interfaces*, **2021**, 8 (13), 1-14. https://doi.org/10.1002/admi.202100384.

Parchovianský, M.; Parchovianská, I.; Švančárek, P.; Medveď, D.; Lenz-Leite, M.; Motz, G.; Galusek, D. **High-Temperature Oxidation Resistance of PDC Coatings in Synthetic Air and Water vapor Atmospheres**. *Molecules*, **2021**, *26* (8), 2388. https://doi.org/10.3390/molecules26082388.

Petríková, I.; Parchovianský, M.; Švančárek, P.; Lenz Leite, M.; Motz, G.; Galusek, D. **Passive Filler Loaded Polysilazane-derived Glass/Ceramic Coating System Applied to AISI 441 Stainless Steel, Part 1: Processing and Characterization**. *Int. J. Appl. Ceram. Technol.*, **2020**, *17* (3), 998–1009. https://doi.org/10.1111/ijac.13417.

Parchovianský, M.; Petríková, I.; Švančárek, P.; Lenz Leite, M.; Motz, G.; Galusek, D. **Passive Filler Loaded Polysilazane-derived Glass/Ceramic Coating System Applied to AISI 441 Stainless Steel, Part 2: Oxidation Behavior in Synthetic Air**. *Int. J. Appl. Ceram. Technol.*, **2020**, *17* (4), 1675–1687. https://doi.org/10.1111/ijac.13531.

Justus, T.; Gonçalves, P.; Seifert, M.; Leite, M.; Probst, S.; Binder, C.; Motz, G.; Klein, A. **Oxidation Resistance and Microstructure Evaluation of a Polymer Derived Ceramic (PDC) Composite Coating Applied onto Sintered Steel**. *Materials (Basel).*, **2019**, *12* (6), 914. https://doi.org/10.3390/ma12060914.

Lenz Leite, M.; Barroso, G.; Parchovianský, M.; Galusek, D.; Ionescu, E.; Krenkel, W.; Motz, G. **Synthesis and Characterization of Yttrium and Ytterbium Silicates from Their Oxides and an Oligosilazane by the PDC Route for Coating Applications to Protect Si_3N_4 in Hot Gas Environments**. *J. Eur. Ceram. Soc.*, **2017**, *37* (16), 5177–5191. https://doi.org/10.1016/j.jeurceramsoc.2017.04.034.

Furtat, P.; Lenz Leite, M.; Ionescu, E.; Machado, R. A. F.; Motz, G. **Synthesis of Fluorine-Modified Polysilazanes via Si–H Bond Activation and Their Application as Protective Hydrophobic Coatings**. *J. Mater. Chem. A*, **2017**, *5* (48), 25509–25521. https://doi.org/10.1039/C7TA07687H.

Seifert, M.; Lenz Leite, M.; Motz, G. **Formation of Mg-Silicates by Reaction of Perhydropolysilazane with MgO during Pyrolysis in Air**. *J. Ceram. Soc. Japan*, **2016**, *124* (10), 1003–1005. https://doi.org/10.2109/jcersj2.16026.

10.6 Curriculum Vitae

Personal Information

Name:	Mateus Lenz Leite
Address:	Königsallee 6, 95448 Bayreuth, Germany
Date of birth:	11.02.1992
Place of birth:	Itajaí, Brazil
Marital status:	Single

Education

2009	Colégio Energia Balneário Camboriú, Brazil
Feb. 2010 - Aug. 2015	Undergraduate studies in Chemical Engineering at the Federal University of Santa Catarina Florianópolis, Brazil
Sept. 2012 - Sept. 2013	Academic exchange in Chemical Engineering at the University of Waterloo Waterloo, Canada
Jan. 2015 - July 2015	Academic exchange in Materials Science at the University of Bayreuth Bayreuth, Germany

Professional Experience

March 2016 - Dec. 2021	Research assistant at the Chair of Ceramic Materials Engineering (CME) at the University of Bayreuth Bayreuth, Germany

www.ingramcontent.com/pod-product-compliance
Ingram Content Group UK Ltd.
Pitfield, Milton Keynes, MK11 3LW, UK
UKHW022000190726
13853UKWH00004B/1635

9 783736 975828